THE COMPLETE ILLUSTRATED HISTORY OF
WORLD WAR II

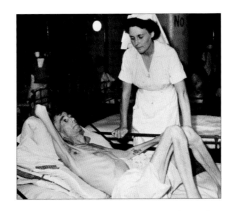

THE COMPLETE ILLUSTRATED HISTORY OF
WORLD WAR II

AN AUTHORITATIVE ACCOUNT OF ONE OF THE DEADLIEST CONFLICTS IN HUMAN
HISTORY WITH ANALYSIS OF DECISIVE ENCOUNTERS AND LANDMARK ENGAGEMENTS

DONALD SOMMERVILLE

LORENZ BOOKS

Contents

Introduction

Within a few years of its end many people even in the victorious countries were questioning if World War I had been a just and necessary war. This has never been true of World War II, either in the immediate aftermath or in longer historical retrospect. The unmitigated evil at the heart of Hitler's Germany and the unrestrained cruelty of the Japanese regime to its prisoners and subjects were both so plain that, at the time and since, few have argued that it was a war that was not worth fighting. Although the world soon moved into the Cold War, an era of potentially even more dangerous confrontation, few have ever suggested that the Allied lives lost in the war were lives sacrificed in vain.

Far more than any previous major war, WWII saw civilians effectively in the front line. In part this arose from the murderous nature of the totalitarian regimes of Hitler and Stalin, and the vicious racism of Japan's militarists, but the Anglo-American bombing campaigns meant that those countries' leaders also had far from clean hands.

Unlike World War I – the first war of the combat aircraft, the tank and the submarine – there were no new types of weapon of any importance introduced until almost the end of the war. The major military development was the extension of the use of air power to a level far beyond anything previously attempted or contemplated. Developments in radio and electronics also meant that, both in the air and on the ground, operations could be directed with a new level of sophistication – tank commanders could talk to each other; aircraft could be detected hundreds of miles away; and radar could find the periscope of a submerged submarine at sea.

The war's one new weapon was obviously the atom bomb, which clearly changed the nature of warfare for all time. Its level of destructive power was so great that even when no nuclear retaliation was possible, its owners might hesitate to use it. With such a dire threat to the future of humankind in existence, it was in a sense just as well that people had seen the flattened cityscape of

Below: London's St Paul's Cathedral, surrounded by fire during the Blitz, autumn 1940. St Paul's survived with little damage, a symbol of Britain's defiant resistance.

Below: Soviet infantry climb out of their trench to attack nearby German positions during the vicious close-quarter battle for Stalingrad in the late autumn of 1942.

Hiroshima and heard the dreadful evidence of the horrors of the Holocaust. The propagation of such knowledge of just how terrible wars can be may have served to make people and governments more likely to resolve their differences peacefully. This is perhaps the lasting benefit that World War II brought.

HOW THIS BOOK IS ORGANIZED

This work has five main chapters, one each on specific phases of the European and Pacific Wars, plus an introductory chapter on the causes of the war and a concluding chapter on its aftermath and long-term legacy. Each chapter is split into separate two-page sections that cover the war's battles and campaigns or the weapons that were used. In addition, there are feature boxes on various subjects, including key personalities and points of special interest. Added together, all these elements provide a detailed, highly illustrated history of World War II that explains every stage of the war's progress, and shows how it affected different parts of the world.

Above: Emaciated survivors of the newly liberated Buchenwald concentration camp talk to some of the American troops who freed the camp, 18 April 1945.

Below: Hitler Youth fighters who have knocked out Soviet tanks are decorated by their leader outside the Berlin bunker where he would commit suicide days later.

Timeline

World War II was truly a global conflict, with inter-related campaigns being fought simultaneously on land and in the air across vast areas of Europe, Asia and Africa and in every ocean.

1931–38

INTERNATIONAL EVENTS Japan begins occupation of Manchuria (Sept 1931); Hitler becomes German Chancellor (Jan 1933); Italy invades Abyssinia (Oct 1935); Germany occupies Rhineland (Mar 1936); Italy annexes Abyssinia (May 1936); Japan invades China (July 1937); "Rape of Nanking" (Dec 1937); Germany annexes Austria (Mar 1938); Germany begins occupation of Czechoslovakia (Oct 1938)

1939

INTERNATIONAL EVENTS Britain and France ally with Poland (Mar); Nazi–Soviet Pact agreed (23 Aug); Britain, France, Australia, India and New Zealand declare war on Germany (3 Sept); South Africa declares war on Germany (6 Sept); Canada declares war on Germany (10 Sept); US "Cash and Carry" law starts (Nov)
POLISH CAMPAIGN Germany invades Poland (1 Sept); USSR invades Poland (17 Sept); Warsaw captured (27 Sept); last Polish troops surrender (3 Oct)
RUSSO-FINNISH WAR USSR invades Finland (30 Nov)

1940

INTERNATIONAL EVENTS Italy declares war on France and UK (10 June); Baltic States seized by USSR (July); Italy, Germany and Japan agree Tripartite Pact (Sept); UK–USA "destroyers for bases" deal (Sept)
RUSSO-FINNISH WAR Ends with partial Soviet victory (12 Mar)
NORWAY AND DENMARK Germany invades (9 Apr); Allied resistance ends (9 June)
WESTERN FRONT Germany invades France and Low Countries (10 May); Dunkirk evacuation (26 May–4 June); French surrender (22 June)
BRITAIN Winston Churchill becomes Prime Minister (10 May); Battle of Britain (Aug–Sept); Blitz against British cities (Sept–May 1941)
UNITED STATES Conscription introduced (Sept); President Roosevelt re-elected (Nov)
BALKANS Italy invades Greece from Albania (Oct)
NORTH AFRICA Italy invades Egypt (Sept); British counter-attack begins (9 Dec)

1941

INTERNATIONAL EVENTS Lend-Lease Act becomes law in USA (Mar); Japanese–Soviet neutrality agreement (Apr); Romania and Italy declare war on USSR (22 June); Japan's assets in USA frozen (July); USA and UK agree Atlantic Charter (12 Aug); UK declares war on Finland, Hungary and Romania; UK, USA and many other nations declare war on Japan (8 Dec); Germany and Italy declare war on USA (11 Dec)
BALKANS Germans conquer Yugoslavia (6–17 Apr); Germans conquer mainland Greece (6–30 Apr); Germans invade Crete (20 May–1 June)
NORTH AFRICA German troops arrive (Feb); British Crusader offensive begins (18 Nov)
EASTERN FRONT Germany invades USSR (22 June); Smolensk pocket eliminated (5 Aug); Kiev encirclement complete (19 Sept); German attack on Moscow starts (2 Oct); Moscow counter-offensive begins (5 Dec); Hitler takes over as Commander-in-Chief of German Army (19 Dec)
PACIFIC WAR Japan attacks Pearl Harbor (7 Dec); Japan attacks Malaya and Philippines (8 Dec)

1942

INTERNATIONAL EVENTS United Nations Declaration issued by Allies (1 Jan)

HOLOCAUST Nazi Wannsee Conference plans "Final Solution" (20 Jan)
PACIFIC WAR Japanese capture Manila (2 Jan); Japanese capture Singapore (15 Feb); US troops surrender in Philippines (8 Apr); Battle of Coral Sea (7–9 May); Battle of Midway (4–6 June); US landings on Guadalcanal (7 Aug)
EASTERN FRONT Battle of Kharkov (15–27 May); Battle of Stalingrad begins (mid-Sept); Soviet counter-offensive near Stalingrad (19 Nov)
NORTH AFRICA Rommel attacks Gazala Line (26 May); Battle of El Alamein (23 Oct–4 Nov); Operation Torch begins (8 Nov)

1943

INTERNATIONAL EVENTS Casablanca Conference, Allies announce unconditional surrender policy (14–24 Jan); Italy surrenders to Allies (8 Sept); Italy declares war on Germany (13 Oct); UK, USSR and USA meet in Teheran Conference (Nov–Dec)
PACIFIC WAR Allied landings on New Georgia (20 June); US landings in Gilbert Islands (20 Nov)
EASTERN FRONT German surrender at Stalingrad (2 Feb); Battle of Kursk begins (5 July); Kiev falls to Soviets (6 Nov)

NORTH AFRICA Battle of Kasserine (14–24 Feb); Axis surrender in Tunisia (13 May)
ITALY Invasion of Sicily (10 July); capture of Sicily complete (17 Aug); Salerno landings (9 Sept)

1944

INTERNATIONAL EVENTS Romania agrees armistice with Allies, declares war on Germany (23 & 25 Aug); Bulgaria surrenders (8 Sept); Finland agrees armistice (10 Sept)
EASTERN FRONT End of siege of Leningrad (27 Jan); Soviets capture Minsk (3 July); Warsaw Rising (1 Aug–2 Oct); Partisans liberate Belgrade (20 Oct)
MEDITERRANEAN Allied landings at Anzio (22 Jan); Rome liberated (4 June); British liberate Athens (14 Oct)
WESTERN FRONT D-Day (6 June); Normandy breakout (1 Aug); landings in southern France (15 Aug); Allies capture Antwerp (4 Sept); Operation Market-Garden (17–26 Sept); Battle of Bulge begins (16 Dec)
PACIFIC Battle of Philippine Sea (19–20 June); landings on Leyte (20 Oct); Battle of Leyte Gulf (24–6 Oct)
BURMA Battles of Imphal and Kohima (Mar–July)

1945

INTERNATIONAL EVENTS Roosevelt dies, Truman becomes President of USA (12 Apr); Hitler commits suicide (30 Apr); Germans surrender in: Italy (2 May), north Germany (4 May), sign overall surrender (7 May); VE-Day (8 May); Potsdam Conference (17 July–2 Aug); Attlee becomes UK Prime Minister (26 July); USSR declares war on Japan (8 Aug); Japan agrees surrender (14 Aug); VJ-Day (15 Aug); Japan signs surrender (2 Sept)
WESTERN FRONT Allies cross Rhine at Remagen (7 Mar); US and Soviet troops meet on Elbe (25 Apr)
EASTERN FRONT Soviets take Warsaw (17 Jan); Vienna falls to Soviets (13 Apr); fighting ends in Berlin (2 May); last German troops surrender in Czechoslovakia (13 May)
BURMA British capture Mandalay (20 Mar); British capture Rangoon (3 May)
PACIFIC Landings on Luzon (9 Jan); landings on Iwo Jima (19 Feb); landings on Okinawa (1 Apr); A-bomb attack on Hiroshima (6 Aug); Soviet attack in Manchuria and A-bomb attack on Nagasaki (9 Aug)

Troubled Times: USA and USSR Soviet paratroop training during the 1930s.

Failing Economies A disabled war veteran begging in the streets of Berlin in the 1920s.

Fascism and Nazism Horst Wessel, later lauded as a martyr in the Nazi cause, on parade in 1929.

APPROACH TO WAR

World War I – the "war to end all wars" – had been centred in Europe but in fact it solved few of Europe's problems. Indeed, it established other problems that the dictators and militarists of the 1930s would exploit to achieve their ends, or so they hoped – but instead this went on to create a new and more terrible conflict.

WWI had been the most costly conflict in human history, both in terms of loss of life and in physical destruction. Not only the killing but the whole process of war production had been thoroughly mechanized and industrialized. Previous standards of humane conduct in types of weapons used and the involvement of civilians were also quickly disregarded. World War II would take all these trends to new extremes.

Germany's ability to dominate Europe had not been removed by WWI. The public will for this to happen remained a force in German life, not just for Hitler and his fellow fascists. Similarly, Japan's leaders, victorious Allies in WWI, came to feel that they had been denied their just rewards and that their national destiny was being circumscribed by racist Europeans and Americans. In these tough economic times, other nations failed to respond well to such issues. Some chose isolationism, others negotiation – and its feeble cousin appeasement – backed by a half-hearted rearmament. When faced with enemies who were happy to start wars whenever it suited them, these strategies were inadequate.

Troubled Times: Western Europe Nationalist troops in Madrid during the Spanish Civil War.

Fascism and Nazism Mussolini (wearing red sash) seized power in Italy in 1922.

Troubled Times: USA and USSR Josef Stalin's purges gravely weakened the USSR in the 1930s.

The Legacy of World War I

The peace settlement that ended the Great War stored up many problems for the future. The former Anglo-French Commander-in-Chief, Marshal Ferdinand Foch, made the best assessment: "This is not a treaty. It is an armistice for twenty years."

The victorious Allies came to Paris for the negotiation of the treaties to end World War I determined to prevent a repetition of the conflict. However, the steps they took to achieve this proved to be ill-judged. Separate treaties were agreed with each of the former Central Powers (usually referred to as the Treaty of Versailles, however, in fact this was the one concluded with only Germany). The three most powerful victors (the USA, Britain and France) dominated the negotiations, though representatives from many other Allied countries were also present.

However, the leaders of the new USSR and the old Russia were still fighting a destructive and divisive civil war and were not included in talks. Also absent were representatives of

Below: Anti-communist White Russian cavalry in action during the Russian Civil War of 1919–21.

Above: The signing of the Treaty of Versailles in the Hall of Mirrors at the palace of Versailles.

the defeated Central Powers; they would be summoned to sign once the text of the treaties had been agreed but had no say in negotiations. These absences were the source of the first great weakness of the Versailles settlement: in Germany, in particular, it would be characterized in

years to come as the "Dictate of Versailles" and therefore by definition was rejected as unjust.

A NEW EUROPE

The treaties redrew the map of Europe, which again stored up trouble for the future. The Austro-Hungarian Empire was cut apart: Austria and Hungary became separate nations and two new nations, Czechoslovakia and Yugoslavia, were founded. Romania took territory from Hungary and Bulgaria, while the Baltic States became independent, largely being created from the Russian Empire. So, too, was Poland, which also included a former German province and, more controversially, gained access to the sea via the "Polish Corridor" at Danzig. This split the main part of Germany from the province of East Prussia. France regained Alsace and Lorraine, taken by Germany after the Franco-Prussian War

of 1870–1. Belgium and even Denmark also gained former German territory. In all Germany lost 13 per cent of its pre-war area, while its colonies became possessions of one or other of the Allies.

Germany's armed forces were to be reduced to a fraction of their previous size. To add yet more resentment, Germany had to accept a clause stating that the war had been entirely its fault (which was far from true) and that Germany must therefore pay substantial financial reparations for the damage and destruction caused. The size of the bill was to be fixed later.

GROWING RESENTMENT

There were many problems with all this, not just in the anger stored up in Germany at what were in many ways real injustices. Almost none of the new nations had an ethnically homogeneous population and most were unhappy with where their borders had been set. Although Britain and France were obviously the most powerful European nations, having

Above: The German battle-cruiser *Hindenburg* scuttled by its crew in protest over the peace negotiations.

suffered so much in the war and with problems of their own they were unwilling to take the lead in making the settlement work.

To perform such a role a new international body, the League of Nations, was set up, largely at the urging of the USA's President Woodrow Wilson. However, it remained to be

seen how well it would function and, sadly, from the beginning it was limited by the refusal of the US Congress to allow the USA to join.

None of this made World War II inevitable, or anything like it, but anger at the way Germany had been treated did inspire an ex-corporal called Adolf Hitler to join a tiny radical political group named the German Workers' Party in Munich in September 1919.

THE WASHINGTON NAVAL TREATY

Japan, one of the Allied powers in WWI, felt that it had not been adequately rewarded for its efforts because the Versailles Treaty granted it little more than a handful of former German islands in the Pacific. By the early 1920s a naval arms race had developed, with Japan, the USA and Britain all making plans for fleets of new and huge battleships, which in truth no one really wanted or could afford. In February 1922 these three, along with France and Italy, agreed the Washington Treaty to limit their future naval forces. The Japanese, in particular, were reluctant signatories and would come to believe that, yet again, they had been swindled of their just status by a racist conspiracy of Anglo-Saxons. Although the Treaty and later successors would remain in force until the late 1930s, this became another source of trouble for the future.

Below: French troops occupying Essen in the Rhineland in 1919.

Failing Economies

The world economy had been left unbalanced by the debts owed both by the victors and the losers after WWI, and national leaders, with little idea how to manage their domestic economies, were unable to respond effectively to the problems this caused.

The years immediately after WWI saw many hardships and political problems for people all around the world, not just in defeated Germany. Both the Nazi Party and Italian Fascism had their origins in this period. Then, after a period of growing prosperity and seeming progress toward international amity, the collapse of American share prices in 1929 ushered in the Great Depression and an era of political turmoil.

In all the former warring countries, ex-soldiers had to find their way back into civil society but did not always find jobs easy to come by; the many disabled veterans found life particularly difficult.

Poverty was an everyday reality for millions of war widows and orphans. Those higher up the social scale worried that the

Above: A German housewife uses worthless paper money to light her stove during the 1923 crisis.

Below: Having left Germany when the Versailles Treaty was signed, French troops returned in 1923 when reparations payments were delayed.

turmoil of Russia's communist revolution might spread. In Germany right-wingers formed militias to fight socialists and revolutionaries, and in Britain and other countries troops clashed in the streets with strikers and other protesters.

With its economic struggles, Germany fell behind in its reparations payments. In retaliation French and Belgian troops occupied the Ruhr in January 1923, an action that triggered a financial crisis and massive inflation in Germany. By late 1923 the mark's foreign exchange value was 130 billion to the dollar. In Germany money was virtually worthless paper and many in the middle classes saw their savings wiped out.

AMERICAN LOANS

In 1924 an Allied committee, which was led by American banker Charles Dawes (shortly to become US Vice-President), set up a new plan for Germany's reparations payments, which brought stability and growth to the German economy (but only on the back of US loans). Other countries were also heavily indebted to the USA for loans raised during WWI; they, too, struggled to repay these because of a decline in international trade, partly caused by the USA's protectionist measures.

For the moment, however, the American economy and stock market were booming, while in international affairs

Above: Fooled by steadily rising returns through the 1920s, many Americans speculated on the stock market and had to pay the price.

there seemed to be other promising signs. In 1926 Germany joined the League of Nations and seemed to be reasonably stable in its domestic politics, even though radical parties of both left and right had significant followings.

In 1928 a range of leading nations even signed a treaty – called the Kellogg–Briand Pact (after the US secretary of state and the French foreign minister who led the negotiations) – by which they agreed to renounce the use of aggressive war as an instrument of national policy. Obviously, this agreement did not fulfil its aims, but

it has remained important in international law. It would be used after WWII to support a number of the charges made against the leaders of Germany and Japan by the International War Crimes Tribunals.

Below: A street scene in the early 1920s, in fact in Berlin but with equivalents in many places.

In 1929 a plan proposed by Owen Young, another American banker, finally fixed the amount of war reparations Germany still had to pay. Though the terms were milder than those previously in force, many Germans still believed they were overly harsh, even if they allowed this issue finally to be closed.

ECONOMIC COLLAPSE
Then, in October 1929, prices collapsed on New York's Stock Exchange and the credit that had financed the USA's own economic growth and kept many foreign economies healthy dried up. Soon there would be millions of unemployed in all industrialized nations and great opportunities for politicians offering radical solutions to these problems.

Fascism and Nazism

In Italy, and then in Germany, radical political movements with dominating leaders took control of the governments and then the whole of national life, remaking these countries in a new, violent, intolerant and ultra-nationalist form.

Both Italy and Germany saw much use of violence to suppress strikes and socialist activities immediately after WWI. In Italy the National Fascist Party, which was led by journalist and war veteran Benito Mussolini, was prominent in these activities. Backed by numerous party thugs (*squadristi*, or Blackshirts), Mussolini took advantage of continuing political instability to have himself appointed Prime Minister in 1922. Then, in 1924–5, a further crisis (caused by the Blackshirt murder of a leading socialist) saw Mussolini discard the pretence of constitutional rule and begin the process of making Italy a fascist state.

Below: A Nazi parade in 1929, led by Horst Wessel, later celebrated as a martyr in the Nazi cause.

Above: Mussolini (wearing sash), during the 1922 "March on Rome" when he seized power in Italy.

FASCIST RULE

Mussolini's message was ultra-nationalist and stated that the good of the nation came before any individual rights or liberties. Elections were first rigged and then abolished, as were political parties and freedom of speech; fascist organizations were set up in every walk of life, starting with one for boys of six and upward. Mussolini himself became *Il Duce* ("the leader") and took unchallenged control of the government, supported by propaganda that claimed, among other things: "Mussolini is always right."

Hitler's rise to power in Germany followed a similar though rather longer route. After building his tiny political party, renamed the National Socialist German Workers' Party, into a significant local force, Hitler mounted an attempted coup in Munich in 1923. This "Beer Hall Putsch" failed and Hitler was jailed for a time. However, he did gain a national reputation from this event – one of his co-conspirators was General Erich Ludendorff, one of Germany's most famous soldiers in WWI.

In jail Hitler wrote his political manifesto *Mein Kampf* (usually translated as "My Struggle"), which set out very clearly the core of his beliefs: a hatred of Jews and communists and an intention to gain new territory for the German race in eastern Europe (*Lebensraum*, or "living space").

NAZIS IN POWER

Hitler and his party remained fairly minor players until the world economic crash began in 1929. Now allied with big business interests and preaching that Germany's ills were caused

ADOLF HITLER

Hitler (1889–1945) alone did not cause WWII, but he did more than anyone else to bring it about. His ideas and plans shaped the war's character, helping make it the most brutal conflict in human history. Hitler, an Austrian by birth, fought as a German soldier in WWI and went into politics shortly after, disgusted by the outcome of the war. His political views were based on anti-communism and anti-semitism, and a sense of his own and Germany's destiny to rule. His regime was chaotic and, as time went by, his orders were increasingly irrational. He was as responsible for Germany's ruinous defeat in 1945, as he had been for the astonishing successes of the early war years.

Right: Hitler and members of his cabinet following his appointment as chancellor in January 1933.

from abroad and by the detested Treaty of Versailles, Hitler found substantial new support.

In 1932 the Nazis became the largest party in the Reichstag (Germany's parliament). After further elections and some complicated political manoeuvring, Hitler was then appointed as chancellor at the head of a multi-party government by President Hindenburg in January 1933.

By late 1934, after Hindenburg's death, Hitler had taken complete control and ruled with legally acquired emergency powers. He became the *Führer* ("leader"), combining the offices of president and chancellor; all opposition political parties were banned; concentration camps, the Gestapo secret police and the rest of the apparatus of a police state were established; and the members of Germany's armed forces all swore a loyalty oath to Hitler personally.

In 1934 Hitler also confirmed his own supremacy within the Nazi movement in the "Night of the Long Knives". The party's left-wingers and the leader of the party militia (the SA or Brownshirts) were among those killed. The SS, led by Heinrich Himmler, became more important. Legal persecu-

tion of Germany's Jews began in earnest with the Nuremberg Laws of 1935, which withdrew many civil rights. At the same time Jews suffered increasingly from thuggery and intimidation.

Below: Nazi thugs making an example of "criminals" – a Jew and his Christian girlfriend.

Aggression: Europe and Asia

Throughout the 1930s first Japan, then Italy and finally Germany sought territorial gains by violent means. They backed their aggression with a ruthlessness seldom seen in international affairs and with new and powerful armaments.

World War II began, some would say, in September 1931, when Japan attacked and in effect annexed the Chinese province of Manchuria. The attack was inspired by a conspiracy of middle-ranking Japanese Army officers who faked a sabotage incident on the South Manchurian Railway, owned by the Japanese. The government in Tokyo knew nothing of this but went along with events when a full-scale advance was immediately started. Throughout the approach to war both the Japanese Army and Navy (and factions within them) would in effect be laws unto themselves.

In February 1932 Manchuria was declared independent by the Japanese as Manchukuo, but in reality it was a puppet state economically exploited

Above: Japanese troops march into Nanking in December 1937.

under Japanese control. In 1933 when a League of Nations investigation finally stated what was obvious (that Japan was the aggressor), Japan simply left the League and increased its military budget.

THE NAZI THREAT

Hitler also left the League in 1933 – ostensibly because other countries had refused to disarm

to Germany's level, but the reality was different. In 1935 he reintroduced compulsory military service and announced the existence of a German air force, both outright violations of the Treaty of Versailles. Britain, France and, at this stage, Italy condemned these moves but this was a sham. Later the same summer, without prior consultation with France, Britain concluded the Anglo-German Naval Agreement, specifically allowing expansion of the German Navy and the creation of a U-boat force.

Italy was the next to make a move. Aware of the country's modest resources, Mussolini's government had not sought gains abroad in the first decade of fascist rule, though it had brutally consolidated control

Below: German troops crossing the Rhine bridge into Cologne in March 1936, Hitler's first advance.

Below: Puyi, Chinese puppet ruler of Manchukuo, attends a ceremony with his Japanese backers.

over Italy's existing colonies in Africa. However, in October 1935, Italy attacked Abyssinia. By May 1936 it had completed the annexation, despite half-hearted sanctions introduced by the League of Nations.

Encouraged by this, in March 1936 Hitler sent troops into the Rhineland area, demilitarized by the Versailles terms. Again, all Britain and France could muster were brief protests. Hitler made his move despite opposition from his generals. By being proved right, his control over them was made more secure and his belief in his own military judgement confirmed.

In February 1938 he completed his hold over his armed forces by dismissing the two top generals, the war minister and the commander-in-chief, taking over the supreme command himself. In March 1938 the next step was to merge Austria into Germany, sending troops across the border after Austrian Nazis had destabilized the government. Clearly, this was not going to be his last target.

WAR IN CHINA

With the establishment of Manchukuo, there had been various clashes between the Japanese Kwantung Army there and the Chinese across the border in northern China. In July 1937 the Kwantung Army (not under control from Tokyo), used an exchange of shots near the Marco Polo Bridge on the outskirts of Beijing as an excuse for escalation. There was heavy fighting around Shanghai for several months from August, before the Japanese made their move inland to Nanjing (then usually known as Nanking) in

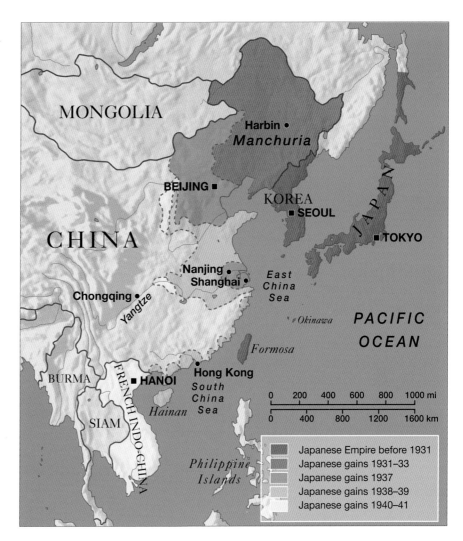

CHINA AND MANCHURIA
Japan's annexation of Manchuria and advance into China up to December 1941.

December. When they captured the city the Japanese forces went on a rampage of murder, looting and rape for several days. This "rape of Nanking" was reported around the world and widely condemned.

Although stronger than it had previously been, Jiang Jieshi's (Chiang Kai-shek's) Nationalist Chinese government could not prevent the Japanese, with more than a million men deployed, taking control of much of northern and eastern China over the next two years.

Above: German tanks parade through Vienna in 1938, after Austria's *Anschluss* with Germany.

Troubled Times: USA and USSR

The 1930s were times of social, political and economic upheaval in the USA and USSR. In their very different ways both countries failed to develop effective responses to the new challenges from Germany and Japan.

In the 1930s the USA and the USSR were not the super-powers they became after WWII. As well as being pre-occupied with economic difficulties, and political controversy over how to respond, the American people were unsure what their country's role in the world ought to be: should they participate fully or were they better off if they stood apart?

For its part the USSR was rent through the decade by a range of self-inflicted wounds as the economy was completely reworked and the rule of Josef Stalin and the Communist Party became yet more brutal and unchallengeable.

THE UNITED STATES

President Franklin Roosevelt was inaugurated in March 1933 with a quarter of the US work-force unemployed and the gross national product half that of the late 1920s. His New Deal poli-cies encompassed a range of measures: of public works, wel-fare and poor relief, reform of the banking system and more. For Roosevelt and the American people the priority for the remainder of the decade would always be domestic concerns.

His achievements delighted some Americans while others despised him. Economists remained unsure how effective many of his policies were, but there was no doubt that he greatly enhanced the power and prestige of the presidency in ways that would become impor-tant in the following decade.

Most Americans agreed that their nation had blundered unwisely into war in 1917 and felt that their country should avoid future conflicts, especially in Europe. The USA had refused to join the League of Nations in 1919. In 1935, and again in 1937, Congress passed neutrality laws to try to ensure

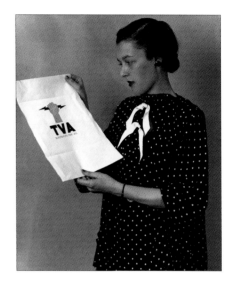

Above: The Tennessee Valley Authority (TVA) was one of Roosevelt's New Deal initiatives.

that the USA was not again dragged into war. In any case the USA had comparatively little military muscle. The US Navy was indeed a world force, allowed by treaty to be equal in size with the British Royal Navy. However, only with re-armament measures at the end of the 1930s did it actually build up to its permitted level. In addition, the US Army was so small as to be internationally irrelevant. Nothing in all this was likely to deter aggression.

THE USSR

Throughout the 1930s the USSR was as inward-looking and in an even deeper state

Left: Strikers and police clash at a Michigan car plant in 1937, a common sight in 1930s' USA.

of turmoil than the USA. The government's policies of mass industrialization and the collectivization of peasant farms were pushed through. On one hand they created the industrial base that could withstand Hitler's attack in 1941, but on the other the cost was huge, with up to 15 million people dead from murder and famine in the rural areas.

All opposition, whether real or imagined, to the rule of the Communist Party was crushed in the so-called Great Terror, with millions of citizens being shipped off to the brutal labour camps of the Siberian GULag. During this terrifying process the state became governed not by the dictatorship of the proletariat, as provided for in Marxist-Leninist theory, but by the absolute rule of one man: Josef Stalin.

The USSR's huge military potential was clear. Soviet forces had pioneered experiments – in paratroop operations, for example – and later in the decade Marshal Mikhail Tukhachevsky was working to create new tank formations, similar to those that would soon be seen in Germany. In addition, the process of industrialization saw both the quantity and quality of military equipment greatly improve by the late 1930s.

Tukhachevsky, however, and most senior commanders as well as about half the junior officers were all executed during the Terror. Their replacements, Stalin's creatures, like all senior officials, were slow to repair the damage done. Consequently, the USSR was not at all well prepared for the coming war.

JOSEF STALIN

Stalin (1879–1953) was the dictator who ruled the USSR with an increasingly tight grip after the death of Lenin in 1924 and who would extend his control to cover all of eastern Europe by the end of the war. He tolerated no rivals or expressions of dissent. His paranoia was so extreme that millions of loyal Soviet citizens were murdered or sent to the GULag slave-labour camps. His cold and cynical calculations led him astray in 1941 when he refused to prepare properly for Hitler's attack, but otherwise his ruthless energy drove his country forward to victory and the establishment of a new European empire.

Below: Stalin, with Molotov (left), foreign minister throughout WWII, and Marshal Voroshilov, a poor general but a favoured crony of his leader.

Below: Three out of five Soviet marshals were executed in Stalin's Great Terror, including Mikhail Tukhachevsky (shown here).

Below: Soviet paratroop experiments. After the Terror generals reverted to traditional infantry-based tactics.

Troubled Times: Western Europe

Britain and France had played the main part in defeating German aggression in WWI, but their casualties and economic and social problems left them with diminished resources and little will to repeat the process.

In the inter-war years many British people saw WWI as an aberration. They thought that in foreign affairs Britain was an imperial and maritime power and should concentrate on those aspects. Sending a mass conscript army to fight and die in a continental war was not something Britain historically had been accustomed to do and should be reluctant to repeat, especially since it became increasingly clear that the Versailles Treaty had not come close to solving Europe's problems. British resources had been plundered to win WWI and in the economic crisis of the 1930s, British governments could see no way of standing up to Hitler

Below: The first meeting, in 1920, of the Council of the League of Nations, the body roughly equivalent to the later UN Security Council.

in Europe, if at the same time they had to protect the British Empire against Japan.

There was also a strong feeling in some quarters that the Versailles settlement had indeed been unjust. Some also felt that a Germany restored to a more realistic international status under Hitler's leadership

Left: Stanley Baldwin, three-times British Prime Minister in the 1920s and '30s, was slow to rearm in the face of Hitler's threats.

could at least be relied on to help keep the communist menace confined to the USSR.

Overall, Britain avoided co-operating closely with France until 1939. This had the dual effect of preventing another "continental commitment" and also preserving relations with the dominion nations of the Empire, which were reluctant to become involved in Europe's troubles once again. Britain's failure to work with France was shown in 1935, for example. In April that year Britain joined France and Italy in a conference at Stresa to condemn Germany's breaches of the Versailles Treaty – then two months later unilaterally and totally hypocritically negotiated the Anglo-German Naval Agreement, which specifically allowed Germany to make further breaches.

FRENCH RELUCTANCE

The human and physical cost of WWI for France had been vast. Even before the war France's population had been overtaken by that of Germany and the French economy was nowhere near as solid. Continued demographic and economic weakness in the inter-war years went hand in hand with political instability. Successive insecure governments formed and re-formed.

22

Left: Rebel Spanish Nationalist troops fight their way into Madrid past buildings wrecked by air attacks, a token of much to come.

France had gone to war in 1914 with a wholeheartedly offensive outlook. However, by the 1930s, the situation had changed and any sort of military response in a

THE MAGINOT LINE

Named after André Maginot, the French minister of war who started its construction in 1930, this was a deep and sophisticated system of fortification all along the Franco-German border. Built in line with the overall defensive nature of French strategy, work on it continued through the decade. Blockhouses, anti-tank defences, strong points, major forts and more were all built at a cost of several billion francs. However, the major weakness was that the defences, strong as they were, did not extend to cover the whole of France's northern border.

crisis was out of the question – unless France could muster overwhelming and enthusiastic international support.

ABYSSINIA AND SPAIN

Italy's invasion of Abyssinia in 1935 gave the first clear demonstration of how enfeebled Britain and France had become. Already Japan had proved the weakness of the League of Nations by simply walking out and carrying on regardless when the annexation of Manchuria was condemned. British- and French-led League sanctions against Italy for attacking Abyssinia were hardly stringent – oil supplies were not stopped and Italian troopships were still allowed to use the Suez Canal. Sanctions were abandoned in less than a year after the Italians had overrun Abyssinia.

Italy and Germany had been brought closer by the opposition their various measures had met, ineffective though it had been. This process was confirmed,

in part, by the next major development: the Spanish Civil War, which began in July 1936.

Both Germany and Italy sent significant help to Spain's right-wing Nationalist rebels. The Soviets sent less valuable help to Spain's Republicans, but France's leftist Popular Front government refused assistance. Britain condemned foreign interference in Spain's affairs but took virtually no action to back their words up.

The war was also important for the lessons major armies drew from it. France and the Soviets concluded that tanks were less effective in battle than some had thought; Germany saw things differently, in particular using Spain as the testing ground for new air-combat and ground-support techniques.

Below: Italian ships pass through the British-controlled Suez Canal, demonstrating the weakness of League of Nations sanctions in 1936.

Europe on the Brink

Prime Minister Chamberlain proclaimed "peace for our time" when he returned from Munich, but the reality was that 1938 and 1939 saw Hitler's relentless expansionism drive Europe into a new and terrible war.

Czechoslovakia had been created by the Versailles Treaty. By 1938 about a third of its population were German speakers who lived principally in the border area known as the Sudetenland. For some time the Sudeten German Party, with Nazi support, had been agitating for union with Germany, and in September 1938 there was a new crisis, provoked by Hitler.

Although the Czechoslovak authorities soon had control of the situation, Britain's Prime Minister Neville Chamberlain decided there needed to be a permanent solution to the problem. It would probably have been wise if Britain and France had stood firm and risked war to help Czechoslovakia: the Czech

Below: Joachim von Ribbentrop, Hitler's foreign minister, signs the Nazi–Soviet Pact. Stalin and Molotov smile in the background.

Above: From left, Chamberlain, Daladier (French premier), Hitler and Mussolini at Munich in 1938.

military and its border defences were very strong and the German armed forces far less prepared for war than was later the case. Instead, although France was resigned to fighting, Chamberlain chose to negotiate.

MUNICH AGREEMENT

Chamberlain flew to Munich to meet Hitler, to negotiate a deal to force Czechoslovakia to cede to Germany areas where the majority of the people were ethnic Germans. Without British support, France had little option but to abandon the Czechs as well. In return Chamberlain had Hitler sign a vague friendship agreement and, on his return, announced that he had secured "peace for our time". He seems to have thought that, fundamentally, Hitler was a reasonable statesman who would keep his word – if that were the case, Czechoslovakia's problems were

a local issue not worth provoking another European war over – a war that Britain could not fight without jeopardizing its Empire in the Far East, which was increasingly under threat from the Japanese.

In early October 1938 Germany moved into the Sudetenland and the remainder of Czechoslovakia was split into three autonomous provinces. Then Poland, followed by Hungary, also grabbed disputed areas over the next weeks. In March 1939 Hitler moved in to complete the destruction of Czechoslovakia.

By that point Britain and France were ready to do a little more so they issued a guarantee of support to Poland, which was clearly Hitler's next target. The international status of the city of Danzig (Gdansk), and the Polish Corridor that divided East Prussia from the main part of Germany, had long been seen as an affront to German nationalism. Now Hitler vehemently demanded their return.

Right: Nazis on parade in Danzig (Gdansk) in 1939. Their banner proclaims "Danzig is a German town".

Right: Nazis on parade in Danzig (Gdansk) in 1939. Their banner proclaims "Danzig is a German town".

NAZI–SOVIET PACT

With Hitler now looking east, the USSR's position became more important. In the summer of 1939 Britain and France half-heartedly began talks in Moscow. However, these made little progress when it became clear that for Stalin any alliance was conditional on having the right to station troops in Poland.

Instead, in mid-August, the world was stunned when the communists and Nazis, formerly implacable enemies, became friends with the signing of the Soviet–German Non-Aggression Pact.

What the rest of the world did not know was that secret parts of this deal provided for Poland to be divided up between Russia and Germany and that the Baltic States and Finland would also lose their independence to Russia.

Earlier that summer Britain and France had finally begun military co-operation planning, and Britain had introduced conscription in May. Even at this last moment Chamberlain was still looking for some compromise concession to Hitler that might avoid war.

When a formal Anglo-Polish alliance treaty was announced on 25 August, Hitler hesitated for a few days, postponing Germany's planned attack on Poland from 26 August to 1 September – but at dawn that morning the invasion began.

GERMAN EXPANSION, 1936–9
Poland and Hungary also joined in the partition of Czechoslovakia.

Above: The announcement of conscription in Britain in May 1939.

German border 1919
German border 1 Sept 1939

Tanks, 1939–42 An early model German Panzer 3 during a river-crossing operation.

Air Combat: Weapons, Tactics and Aces The ball turret of a B-17 Flying Fortress.

Anti-tank Guns, 1939–42 A Soviet 45mm M1937 anti-tank gun coming under artillery fire.

German and Italian territory 1939
German-occupied 1939
German-occupied 1940
Annexed by USSR 1939–1940
German allies by spring 1941
German-occupied Mar–May 1941
Axis-occupied Jun–Dec 1941
Front line 5 Dec 1941

0   100   200   300   400   500 mi
0   200   400   600   800 km

ICELAND

ATLANTIC
OCEAN

NORWAY
SWEDEN
FINLAND

North
Sea

IRELAND

UNITED KINGDOM

DENMARK

Baltic Sea

ESTONIA
LATVIA
LITHUANIA
EAST
PRUSSIA

USSR

NETHER-
LANDS
BELGIUM

GERMANY

POLAND

FRANCE

CZECHOSLOVAKIA

SWITZER-
LAND

AUSTRIA

HUNGARY

ROMANIA

YUGOSLAVIA

Black Sea

Caspian Sea

PORTUGAL

SPAIN

Corsica

Sardinia

ITALY

ALBANIA

BULGARIA

TURKEY

Sicily

GREECE

Mediterranean Sea

CYPRUS
(British)

SYRIA
(French)

IRAQ
(British)

IRA

TUNISIA
(French)

PALESTINE
(British)

TRANS-
JORDAN
(British)

ALGERIA
(French)

Apr–Nov
1941

End Sept
1940

Feb
1941

LIBYA
(Italian)

EGYPT
(British)

North Africa front lines

HITLER'S TRIUMPHS

Throughout the period from September 1939 to the autumn of 1941, there was never a moment when an impartial observer would have bet on anyone apart from Germany winning the war. The parade of German military successes seemed endless, with Poland, Denmark and Norway, France and the Low Countries, the Balkans and, finally, most of European Russia coming under German domination. This rapid succession of victories had made Hitler's domestic position unassailable and convinced him of his own military genius, which would prove to be less infallible, however, in the years to come.

Remarkably, Britain was fighting on, with increasing help from the USA, but it was hard to see how anything Britain could do would really hurt Germany – nor did it seem likely that the Red Army could recover from the crushing defeats recently inflicted on it. Also, by late 1941, the brutal nature of the war had been made plain. Both the Nazis and the Soviets had murdered many thousands of Poles, and the Nazis had continued with mass killings of Jews and others on Soviet soil.

Yet the signs of a turning tide were there to be seen. America was rearming fast and inching closer to joining the war even before Japan's surprise attack on Pearl Harbor, while in that same week carefully husbanded Soviet reserves went into action on the Moscow front and threw back the German advance.

The USA and the European War President Roosevelt's "fireside chats" on the radio were famous.

The British Home Front Child evacuees leaving London in 1939 say goodbye to their parents.

The Fall of France A German Ju 87 Stuka bomber, one of the decisive weapons of Blitzkrieg.

Poland and the Outbreak of War

Bolstered by his pact with Stalin, Hitler no longer felt any need to restrain his aggression. Hitler first, and then Stalin, attacked and quickly conquered Poland while Britain and France did nothing other than declare war on Germany.

The first shots of the European half of WWII were fired at a Polish naval base by the old German coast-defence ship *Schleswig-Holstein* early in the morning of 1 September 1939. Already Hitler's Luftwaffe was screaming in to attack Polish air bases and German troops were surging over the borders. Britain and France responded by sending ultimata to Germany demanding an immediate withdrawal. When there was no response they declared war on the 3rd.

THE POLISH CAMPAIGN

Germany deployed 53 divisions for the attack, including all 6 Panzer (or armoured) divisions then in existence. The Army Commander-in-Chief, General Walther von Brauchitsch, controlled the operation with little interference from Hitler. Poland

Below: A Soviet tank passes a line of German troops in Brest-Litovsk, Poland, late September 1939.

KEY FACTS

PLACE: Poland

DATE: 1–28 September 1939

OUTCOME: Poland is attacked and completely overrun by Germany and then the USSR.

---

had not begun mobilizing its forces until 30 August, so many reservists were still on their way to their units. The Poles had about 23 divisions deployed, with very few tanks. However, they did have a significant cavalry force, though it is untrue (as sometimes has been suggested) that the cavalry were used to charge German tanks at various points in the fighting.

In the air German superiority was even more marked, with the Luftwaffe using about 1,600 modern aircraft against the Polish force of some 500 mostly obsolescent types.

With their superior strength, equipment and training the Germans soon gained the upper hand. Poland's Commander-in-Chief, Marshal Edward Smigly-Rydz, had decided to defend along the borders and his troops were soon thrown out of the exposed positions this plan committed them to. By the middle of the month the Polish forces had been split into isolated small groups, while the Germans were closing in on Warsaw – which they captured, ending the campaign on the 28th after a vicious artillery and air bombardment.

By then Stalin had taken his share of the spoils. Soviet troops crossed into Poland on the 17th, as agreed with Hitler in the Nazi–Soviet Pact; by the end of the month the Soviets had occupied about half the country. The official Soviet line was that

Below: German troops in Poland in September 1939. Few German units were motorized like this one.

GLEIWITZ

As a pretext to justify the attack on Poland, an SS unit faked an attack by Polish troops on a German radio station at Gleiwitz (Gliwice) in Silesia near the Polish border. The "attackers" captured the station late on 31 August, made a brief anti-German broadcast, then left after killing a concentration camp prisoner they had brought with them. This unfortunate person was then displayed to foreign pressmen as a civilian victim of Polish brutality.

Above: German troops crossing the San River in southern Poland, watched by their proud *Führer*.

they were intervening to protect the ethnic Belorussians and Ukrainians, but in fact they were annexing this territory. On the 29th the Germans and Soviets announced a friendship treaty that confirmed the partition of Poland.

THE PHONEY WAR

As well as Britain and France and their various colonies, who were given no choice in the matter, Britain's so-called "old dominions" declared war on Germany in the first days of September. Each, however, had some reservations.

For example, the Australian government declared war and introduced conscription, but only for home defence. Canada declared war only after this had been debated in parliament and also decided not to have conscription. The United States proclaimed its neutrality – there was no doubt at this stage that most Americans wanted to keep out of the war in Europe.

Although they had gone to war on Poland's behalf, Britain and France did nothing to help the Poles. For the moment Hitler was content not to provoke them further – German forces in the West in September were very weak. British government ministers refused to bomb industries in the Ruhr because the factories were private property, while French troops made only tentative forward moves in a small area of the Saar. This "Phoney War" would continue on the Western Front until Germany's attack in 1940.

Below: British families building Anderson shelters as a precaution against German air raids, during the Phoney War.

Tanks, 1939–42

The successes of Germany's early-war campaigns were based on the power of the much-feared Panzer divisions, giving tanks a prominence in military affairs they had never before attained.

Tanks, as well as anti-tank weapons, were the only types of land-warfare weaponry that saw substantial development during WWII. In 1939 the most powerful tanks mainly had a gun in the 37mm (2pdr) class and were protected by up to 40mm (1.58in) armour. By 1942, 75mm (2.95in) weapons and twice as much armour were typical. These figures increased still further later in the war.

EARLY PANZERS

Nazi Germany's first tank, built during the mid-1930s, was the Panzerkampfwagen (PzKpfw) 1, a light two-man design armed only with machine-guns. This was soon joined by the slightly more powerful PzKpfw 2. Both saw combat into 1941. The best tanks in the Panzer divisions of 1939–41 were the next two in the series. The PzKpfw 3 was available in various marks in the early-war period. Up to April 1940 all carried a 3.7cm (1.46in) main gun but this was replaced by first a 42-calibre 5cm (1.97in) gun and then a more powerful 60-calibre 5cm. The PzKpfw 4 was originally conceived as an infantry-support vehicle and hence began life with a short (24-calibre) 7.5cm (2.95in) gun; by 1942 its 7.5cm gun was a version twice as long. For the first two or three years of the war the Germans also used many Czech-built tanks. The PzKpfw 38(*t*) was a powerful design of similar capabilities to the PzKpfw 3 of the time.

The British and French tanks facing this array in 1940 were a mixed bag; most were additionally

MATILDA 2

The Matilda was the only British tank to see combat throughout the war, remaining in use against the Japanese until 1945. Its main European service was in France and North Africa in 1940–1. (The example shown below is in Egypt, December 1940.) Its thick armour and reasonable anti-tank gun power made it a tough opponent then but it was very slow cross-country. Almost 3,000 were built.

WEIGHT: 26.9 tonnes
LENGTH: 5.6m (18ft 5in)
HEIGHT: 2.5m (8ft 3in)
ARMOUR: 78mm (3.1in)
ROAD SPEED: 25kph (15mph)
ARMAMENT: 1 x 2pdr (40mm) +
    1 x machine-gun

Left: M3 Lee tanks and American troops in training in Northern Ireland in 1942. These were some of the first US soldiers to serve in Europe.

hampered by being dispersed in small infantry-support units. Various French designs like the Somua S-35 were well-armed and armoured but were made less efficient in action by their one-man turrets, in which the commander had to make tactical decisions and also load, aim and fire the gun.

Britain had a succession of light, medium and heavy tanks throughout the early-war years. The Light Tank Marks 5 and 6 were armed only with machine-guns – so they were useless for anti-tank combat – but did see extensive service in France and the early North African battles.

The medium (or cruiser) tanks went from the Mark 1, first produced in 1938, to the Mark 6 or Crusader, final versions of which appeared in 1942. All the cruiser tanks had reasonable armour and gun power, but they had appallingly unreliable engines and running gear.

The heavy tanks were better. The Matilda 2 carried the same 2pdr (40mm) gun as most of the cruisers and mounted 78mm (3.1in) armour, which made it very difficult indeed to knock out in its heyday of 1940–1. Its successors, the Valentine and Churchill, were robust and reliable, if under-gunned.

The USA's early-war tanks were the M3 Light Stuart and M3 Medium, known as the Lee in US service and the Grant in the slightly different form used by Britain. The Stuart, with a 37mm (1.46in) gun and 37mm armour, was fast and reliable. Updated as the M5, it was still used extensively in 1944–5. The Lee/Grant was an interim type, a hurried redesign of an earlier model to fit a 75mm

Above: An early-model Panzer 3 with 3.7cm (1.46in) gun during a river-crossing exercise in 1941.

(2.95in) gun, but only in a side sponson rather than a turret. Production ended in 1942.

SUPERIOR SOVIET TYPES

The most impressive Allied tanks were those of the Red Army. The early-war BT-7 was particularly fast and carried a reasonable 45mm (1.77in) gun, but the next generation of Russian tanks were the finest in service anywhere in their time. The heavy KV-1 came first, making its combat debut in the Russo-Finnish War in 1940. Its 76.2mm (3in) gun and 90mm (3.54in) armour outclassed anything the Germans had available in 1941. Even better was the medium T-34, sometimes described as the tank that won the war for the USSR. Fortunately for the Germans, comparatively few of these two designs were available in 1941.

SOMUA S-35

Designed in 1934–5 the Somua S-35 was intended as a fast "cavalry" tank and fulfilled this brief well. Some 300 saw combat during the 1940 Battle of France, including those shown below. The S-35 was well-armed and armoured but mechanically unreliable.

WEIGHT: 20 tonnes
LENGTH: 5.5m (18ft)
HEIGHT: 2.7m (8ft 10in)
HULL ARMOUR: 40mm (1.58in)
ROAD SPEED: 37kph (23mph)
ARMAMENT: 1 x 47mm (1.85in)
  + 1 x machine-gun

The Baltic and Scandinavia

Northern Europe's smaller nations continued to fall to aggression through 1939 and 1940. The USSR attacked Finland in November 1939 and annexed the Baltic States in 1940, while Norway and Denmark were taken by the Nazis that spring.

Stalin's plans for his Western neighbours were first demonstrated in the Soviet-occupied parts of Poland. By the end of October 1939 mass arrests and deportations of "enemies of socialism" had begun and fraudulent elections had appointed assemblies, which dutifully voted for incorporation into the USSR.

At the same time Lithuania, Latvia and Estonia were forced to conclude "friendship" agreements, giving the Soviets the right to station troops in these countries. In June 1940 Stalin activated these treaties and sent his troops in. The same process as in Poland quickly followed: terror, false elections and annexation by the USSR. In June 1940 the USSR annexed the Northern Bukovina and Bessarabia regions (formerly part of Romania) as well. They, too, had no option but to yield to Stalin's ultimatum.

Above: Finnish troops used their mastery of the winter conditions to defeat the initial Soviet attacks.

Below: A German tank and German infantry on the advance early in the Norwegian campaign.

THE WINTER WAR

These moves were part of a process of extending Soviet power and, at the same time, establishing a buffer zone against a possible German attack. Finland had been assigned to the Soviet sphere of interest by the Nazi–Soviet Pact and, in October 1939, Stalin began moves against the Finns.

He demanded that Finland cede territory in the Karelian Isthmus area in the south and in the far north, in return for land to be ceded to Finland in the central part of their border region. The Finns, fearing that this unusual generosity on Stalin's part was the thin end of a very nasty wedge, refused. On 30 November Stalin sent in the Red Army to attack Finland.

Right: A British anti-aircraft battery in Norway. The Germans had complete air superiority.

The Finns had 9 divisions facing 26 Soviet divisions, which also had massive superiority in air support, artillery, tanks and every other material category. Remarkably, the Finnish troops (well-trained and mobile on their skis) completely outfought the Soviets until early February 1940, when Soviet reinforcements under a new commander, Marshal Semyon Timoshenko, began wearing them down. An armistice came into effect on 13 March – the Finns conceded territory similar to that originally demanded.

The war had several repercussions. It was the final nail in the coffin of the League of Nations – the USSR was expelled but took no notice of this severe punishment. It also made clear the hesitancy of Britain and France, who considered the possibility of sending help to the Finns (even at the risk of war with the USSR) but did nothing until far too late.

Part of the reason for helping Finland was that, in crossing Norway to get there, the Anglo-French forces would also be able to interrupt Germany's iron-ore supplies from northern Sweden. The difficulty was that neither of these neutral countries wanted foreign troops on their territory. Finally, the war also highlighted what seemed to be the low quality of the Red Army – Hitler for one took note.

DENMARK AND NORWAY
In the aftermath of the Russo-Finnish War, Hitler became convinced (and correctly) that Britain and France still planned to intervene against his Swedish sources of iron ore by moving through Norway. On 9 April 1940 German forces advanced overland into Denmark and attacked a series of points along the Norwegian coast from the air and sea. Neither country had significant armed forces and the Western Allies were taken completely by surprise. Denmark was conquered within hours. Norway took about two months but the result was never really in doubt from the first days, when Germany established air bases in the country that more than compensated for Britain's superior naval strength.

Small Anglo-French forces arrived to help the Norwegians at various points, but the organization of these expeditions was chaotic. The last ones were withdrawn in early June, in the light of the disasters by then occurring on the Western Front.

Below: The German battle-cruiser *Scharnhorst* firing on the British carrier *Glorious* off Norway in June.

Heavy Cruisers

The heavy cruisers of World War II were the product of an intense arms race during the inter-war period. Japan built particularly large and powerful ships by flagrant violation of treaty limits; other navies strove to keep up.

Since battleship design was very tightly controlled by the Washington Naval Treaty and its successors, much inter-war competition between the leading navies came to be in cruiser construction, whether the heavy cruisers armed with 203mm (8in) guns covered here or lighter 152mm (6in) armed types.

In the inter-war period the "treaty limit" for cruisers was 10,000 tons standard displacement. Japan, Italy and later Germany all flagrantly breached this figure, but other nations generally kept fairly close to it. In fact, the limit of 10,000 tons/8in guns had been arrived at rather arbitrarily. It was actually very difficult to build ships to these figures that also had a reasonable balance of armour protection and engine power.

TREATY CRUISERS

Japan and Italy were the first to build "treaty" heavy cruisers, in the mid-1920s. Japan's *Furutaka* was a relatively modest design with six single 200mm (7.87in) guns, reasonable side armour and 33-knot speed, on an official displacement of 7,000 tons (actually about 2,000 more).

Italy's *Trento* and *Trieste* were equally fast with 8 x 203mm guns. However, although their true displacement was about 11,500 tons, they were still rather flimsily built. The four later Italian Zara-class ships were much better armoured and several knots slower.

Japan's successor ships to the *Furutaka* also substantially breached the treaty limits. The Myoko and Atago classes were all over 13,000 tons, carried 10 x 203mm guns and a heavy torpedo armament, with armour up to 120mm (4.72in) thick.

In comparison with these, British and American inter-war ships looked rather second-rate.

Left: The pocket battleship *Admiral Graf Spee* in mid-1939. The *Graf Spee* was scuttled in December 1939 after the Battle of the River Plate.

Britain built a series of similar ships, collectively known as the County class. They all carried 8 x 8in guns on a displacement very close to 10,000 tons. They had very good range and seakeeping qualities but had very little armour and a high silhouette, which made them rather vulnerable in action. Britain also built two smaller ships with 8in guns in the early 1930s – the *Exeter* and *York* – but like most

SISTER SHIPS: 6, inc. 2 Australian
COMMISSIONED: 1928
DISPLACEMENT: 10,000 tons
SPEED: 32 knots
BELT ARMOUR: 115mm (4.53in)
ARMAMENT: 8 x 8in (203mm) + 10 x 4in (102mm) guns

other navies did not build any more 8in heavy cruisers during the war.

Germany operated within a different set of restrictions, having to conform (until Hitler abrogated it) to the Versailles Treaty. In the late 1920s Germany was allowed to begin work on replacements for old coast-defence ships that had previously been permitted under Versailles.

The replacements were three *Panzerschiffe* (armoured ships), soon termed "pocket battleships" by British commentators, but better described as heavy cruisers, as they were eventually officially designated. They carried two triple 280mm (11in) turrets and could reach 26 knots, but significantly exceeded their announced 10,000-ton displacement (cheating that in fact pre-dated Hitler's regime). Supposedly, they had the gunpower to outmatch any cruiser and the speed to escape almost any battleship. However, these ships proved unsatisfactory, with slow-firing armament and unreliable engines. Three later German 203mm-armed cruisers were also built, again large and formidable vessels but troubled by weak engines.

AMERICAN DESIGNS

The US Navy built several classes of "treaty cruisers". First were two Salt Lake City ships with an unusual arrangement of a twin and triple turret fore and aft, with the triples superfiring over the twins; unsurprisingly, they were somewhat top-heavy.

Subsequent American classes changed to three triple 8in turrets, which proved to be a more sensible arrangement. As well as commissioning numerous 6in cruisers, US industrial power also saw the completion of more than a dozen Baltimore-class ships during the war. These kept the same main armament as earlier 8in vessels but increased displacement to 13,600 tons, to fit in the extra equipment that war experience showed to be necessary.

Below: The Astoria-class USS *Vincennes* in July 1942. It was sunk a few weeks later off Guadalcanal.

The Fall of France

The French Army was traditionally one of Europe's strongest, but in 1940 it was utterly defeated in a few weeks of combat. The disaster was caused by poor morale and weak leadership in the face of a ruthless and well-organized enemy.

Although there was no significant fighting on the Western Front before May 1940, many plans were made. Hitler intended to attack, and eventually he approved a radical plan to avoid becoming bogged down in assaults on France's Maginot Line defences. Instead, secondary forces were to advance into Belgium and the Netherlands, defeating these countries but, at the same time, drawing Anglo-French troops forward to help them. Then the main German force, led by most of the Panzer divisions, would advance through the Ardennes region (despite its unpromising terrain) and cut the Allied front in two.

Allied strategy was defensive, aiming to build strength while waiting for Britain's naval

BLITZKRIEG IN THE WEST
France, the Low Countries and the British expeditionary force were all defeated in less than six weeks.

Above: A Ju 87 Stuka in its attack dive, a terrifying experience for those targeted on the ground below.

blockade of Germany to bite. Their plan, in the event of a German attack, was for the best Allied troops to advance into Belgium and link with both the Belgians and Dutch to present a united front. Unfortunately, in their understandable desperation to remain neutral and do nothing to provoke Hitler, the Belgians and Dutch refused to plan jointly with the Allies, so the whole scheme was never properly worked out.

BALANCE OF FORCES
Overall, counting the Dutch and Belgians, the two sides' land forces were roughly equal in May 1940, with some 140

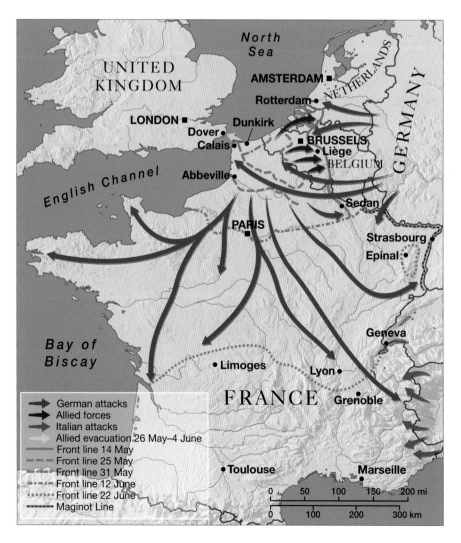

German attacks
Allied forces
Italian attacks
Allied evacuation 26 May–4 June
Front line 14 May
Front line 25 May
Front line 31 May
Front line 12 June
Front line 22 June
Maginot Line

divisions and 3,000 tanks each. However, most of the Allied armour was split up into small scattered groups, while Germany's was concentrated in well-trained Panzer divisions. In the air Germany deployed some 3,000 modern combat aircraft; the Allies had about 2,000, many of them older types. This would be a crucial advantage.

The German offensive began on 10 May. As the Germans had intended, attention was concentrated at first in the north. Within days a combination of heavy air attacks, paratroop operations and advances on the ground smashed the Dutch and Belgian forces. On 14 May the Netherlands surrendered. Britain and France played into German hands – the troops advancing into Belgium were reinforced but they were still unable to hold their positions against the German troops directly attacking them.

CROSSING THE MEUSE
In the meantime, the main German force had reached the River Meuse, crossed it on 14–15 May and was soon racing for the Channel. The Germans reached the coast on the 20th, slicing the Allied armies in two, just as they had planned. Unfortunately, the French Commander-in-Chief had kept no reserves to deal with any such emergency.

Over the next two weeks or so the Allied divisions in the north were forced back into an ever-smaller perimeter around the port of Dunkirk. About 340,000 troops were evacuated from there to Britain by 4 June, including some 120,000 French. The British had to leave all their

BLITZKRIEG

Germany's many successes in 1939–40 were ascribed to a new form of warfare, Blitzkrieg ("lightning war"), a term which is said to have been coined by Hitler himself. It described the combination of fast-moving tank forces and powerful close air support that overran France in a matter of weeks. Although it seemed to depend mainly on armoured units, the secret of Blitzkrieg was more to do with effective co-operation between all the arms of service. This, and the high levels of initiative shown by commanders at all levels, was the true basis of Germany's victories. The Germany Army would remain superior to all its enemies in these areas for most of the war.

equipment behind and many thousands more French troops were taken prisoner.

By then the Germans had regrouped and were ready to finish the job. On the 5th they attacked south from along the

Above: British soldiers crammed aboard a destroyer during the evacuation from Dunkirk.

line of the River Somme and were soon advancing speedily, despite some fierce initial resistance from the remaining French forces. They were in Paris on the 14th and, on the 16th, the French government resigned.

FRANCE SURRENDERS
The new government was headed by Marshal Philippe Pétain. Despite being urged to fight on by the British under their recently appointed Prime Minister, Winston Churchill, Pétain and his cabinet decided to ask for an armistice. France duly surrendered on 22 June.

Below: Panzer 1 light tanks seen during the French campaign. The less capable Panzer 1 and 2 designs were still a significant part of the German force in 1940.

Light Bombers, Recce and Utility Aircraft

Modest armament fits, or even none at all, were the hallmarks of these aircraft types.
Paradoxically, perhaps, the least successful of these designs were the light bombers –
the unarmed reconnaissance and utility types had a far lower casualty rate.

The light bomber category included a number of designs in service in 1939 but most of these were soon found seriously wanting. They were replaced by either the heavier bombers or by purpose-built ground-attack machines.

LIGHT BOMBER DESIGNS

Britain and France both had aircraft of this type in 1939–40. The Fairey Battle had seemed a capable design when it entered service in 1937, but by 1940 its low speed and non-existent armour protection made it, in effect, a deathtrap.

France's Potez 63 series had similar faults and the Breguet 691 was little better, though it did have a more substantial defensive armament. The Bloch

Above: Some 800 examples of the Fw 189 Uhu were used as ground-attack and reconnaissance aircraft.

Below: A Fieseler Storch shows its ability to land (and take off) from unconventional airfields, in this case a Berlin boulevard.

174 was fast (530kph/329mph) and carried a useful 400kg (880lb) bomb load, but only 50 were in service in May 1940. As also to some extent in Britain, the profusion of relatively small French aircraft companies prevented sufficient development

MITSUBISHI Ki-46

The Mitsubishi Ki-46 entered service with the Japanese Army in 1940. It remained in use to 1945, latterly and unsuccessfully as a fighter (shown below), armed with 2 x 20mm (0.79in) cannon in the nose and 1 x 37mm (1.46in) upward-firing gun.

SPEED: 600kph (375mph)
RANGE: 4,000km (2,500 miles)
CREW: 2
ENGINES: 2 x Mitsubishi Ha-102 radials; 1,080hp each
ARMAMENT: 1 x 7.7mm (0.303in) machine-gun

Henschel 123 (built specifically for the attack role). Less capable was the Czech Letov S328, dating back to 1933, still used by some of the Eastern Front's minor air forces in 1944–5.

RECONNAISSANCE

Many well-known aircraft types had variants produced to serve in the reconnaissance role. In the British case specialized versions of both the Spitfire and Mosquito were built for this purpose. Usually, they were unarmed and fitted with uprated engines, along with appropriate cameras for planned high- or low-level missions. The American equivalents included modified Lightning fighters and Havoc bombers. Some designs were given pressurized cabins and other fittings to help them achieve extreme altitudes where they would be very difficult to intercept.

Only one land-based aircraft type was built specifically for the long-range reconnaissance role: Japan's Mitsubishi Ki-46 "Dinah". Over 1,700 of this design were produced and could reach over 600kph (375mph). Range was an impressive 4,000km (2,485 miles).

One of the few aircraft that specialized in the tactical reconnaissance role was Germany's Focke-Wulf 189, which served extensively on the Eastern Front. It was comparatively slow but survived through its toughness and extreme agility.

UTILITY DESIGNS

Most nations had small light transport aircraft, which were also employed for such tasks as artillery spotting and landing agents in enemy territory.

POTEZ 63-11

The Potez 63 family included several bomber, fighter and reconnaissance variants. The 63-11 shown was used mainly in reconnaissance, with about 700 being built, including some for the Vichy air force. Potez 633 bombers served with Romania and Greece.

SPEED: 439kph (273mph)
RANGE: 1,300km (800 miles)
CREW: 3
ENGINES: 2 x Gnome Rhône 14M radials; 700hp each
ARMAMENT: 200kg (440lb) bombs; 3 x machine-guns

These could be found in either army or air force service, according to nationality. General Rommel famously used one such type, a Fieseler Storch, in flights over the North African battlefields, landing from time to time to chivvy on lagging subordinates. Britain's Westland Lysander regularly flew covert missions carrying resistance personnel into France. American equivalents included the Taylorcraft L-2 Grasshopper. The one crucial performance attribute of such aircraft was usually their ability to take off and land in confined and rough areas. None were fast or ever more than lightly armed.

and production effort being given in the pre-war period to the best designs.

Early-war Soviet Sukhoi 2s had similar performance to the above Anglo-French types, with the advantage of reasonable armour protection for the crew. However, many were still shot down by the superior Luftwaffe fighters of 1941.

Although these "modern" designs proved short-lived, some seemingly less capable aircraft (many of them biplanes) fought on in night harassment and similar roles. Aircraft in this category included the Soviet Polikarpov I-153 (originally designed as a fighter) and the German

The Battle of Britain and the Blitz

In the summer of 1940 Hitler seemed unbeatable, but his failure to finish Britain off can be seen as the point when the war changed from being a short one, which Germany would win, to a longer one which the Nazis could conceivably lose.

With France beaten and the British Army practically disarmed after the evacuation from Dunkirk, Hitler probably expected Britain to surrender. However, inspired by Churchill, Britain seemed ready to fight on. On 16 July Hitler therefore ordered his armed forces to start preparing for an invasion of England. Already the Luftwaffe had begun attacks on British shipping in the English Channel, in order to draw the Royal Air Force (RAF) into battle. Since Britain's Royal Navy was still very powerful and much of the German Navy had been lost during the Norwegian campaign, winning air superiority was an essential prelude to invasion.

Above: Firefighters at work on burning buildings in the City of London during the 1940 Blitz.

BRITAIN'S DEFENCES

Fortunately, Britain had made effective preparations. RAF Fighter Command had about 900 Spitfire and Hurricane aircraft, with plenty more being produced to replace the inevitable losses, but trained pilots were in much shorter supply. The defensive organization was excellent, with information from the radar system being fed to a network of control stations and then being used to direct the fighters into combat.

No other nation had such an integrated organization at this time. The Germans did not realize how well this system worked and, accordingly, would not make enough effort to disrupt it by attacks on the radar and control stations.

The Germans had a similar number of single-engined fighters (Messerschmitt Bf 109s) to the RAF, along with over 1,200 twin-engined medium bombers. In addition, they had twin-engined Messerschmitt 110 fighters and Junkers 87 Stuka dive-bombers, about 300 of each; both these types would prove less effective in this campaign than previously.

Whereas Britain was using a defensive system that had been in preparation for many months, Germany had to fight a new sort of battle, very different from their earlier campaigns in support of a land offensive. Their intelligence on RAF strength was very poor, so it was difficult to devise effective plans.

Left: Hawker Hurricane fighters of a squadron with refugee Czech pilots taking off in late 1940.

Right: A formation of Heinkel 111 bombers assembling for a raid during the Battle of Britain.

Above: Air Marshal Hugh Dowding, the successful Commander-in-Chief of RAF Fighter Command during the Battle of Britain.

All-out attacks began in mid-August 1940. There were heavy losses on both sides but the RAF had the advantage initially. Then, for a few days in late August and early September, the Luftwaffe changed its tactics and stepped up strikes against the front-line RAF airfields. This change stretched the RAF to the utmost but the German commanders did not realize that they were winning the battle.

On 7 September Germany switched tactics again and began a series of mass day and night attacks on London, which were heavily defeated on the 15th. On the 17th Hitler postponed his invasion plans. Daylight attacks and air battles continued for several weeks, but the Battle of Britain had been won.

THE BLITZ

London's ordeal was not over. The German bombers came back almost every night, up to 400 or more strong, until late November. By then they were also attacking a range of major cities that included Coventry, Birmingham and others. From November through to May 1941, when the attacks ended because most of the Luftwaffe was being transferred to eastern Europe, the main targets were various port areas like Mersey-side and Clydeside. The British people called these attacks the "Blitz" (from the German word "Blitzkrieg").

At first the British defences were very ineffective. There were few anti-aircraft guns in service and radar-equipped night fighters were only just being developed. Although matters improved as the battle went on, the German loss rate remained low. Some 43,000 British civilians were killed and tens of thousands made home-less in the Blitz, but Britain's war effort was scarcely scratched.

Fighters, 1939–42

In 1940 in the Battle of Britain, the fate of the world depended to a significant degree on the qualities of the two sides' Spitfire, Hurricane and Messerschmitt fighters. Air combat superiority was vital in this and every other campaign.

Like every other kind of military technology, fighters had to have a balance of usually conflicting qualities: speed, rate of climb, manoeuvrability, range, armament, protection and others. Most fighters in service throughout the war were single-piston-engined, pilot-only, low-wing monoplanes. A few biplanes and twin-engined monoplane designs were also produced but generally saw little combat as day fighters.

BIPLANES

Some countries with good modern designs still had a number of biplanes in action in 1939: examples being Germany's Heinkel 51 and Britain's Gloster Gladiator. Typically, they were slow and lightly armed – 400kph (250mph) and four rifle-calibre machine-guns for the Gladiator – and came off badly if facing monoplane opponents. Italy, however, had quite significant numbers of Fiat CR 32 and CR 42 biplanes, and some even continued in use until Italy's surrender in 1943.

MONOPLANES

The classic designs of the era were the Spitfire, Hurricane and Messerschmitt Bf 109. All first flew in 1935–6 and would continue in combat service, albeit in greatly modified forms, until the end of the war. The Spitfire 1 was slightly faster than the Bf 109E (the main versions in service in 1940) and the Hurricane slower than both the others. Both British fighters were more manoeuvrable than the Bf 109, but their 8 x 0.303in (7.7mm) machine-guns were less effective than the Messerschmitt's 2 x 20mm (0.79in) cannon and 2 x 7.92mm (0.312in) machine-guns. Messerschmitts also had a better rate of climb.

Below: A USAAF Curtiss P-40. Many of the 13,700 P-40s built served in the war against Japan.

In 1941 the successor 109F was superior to the Spitfire 5, a balance redressed by the later Spitfire 9. Ultimately, though, there was sufficiently little to choose between them that encounters were more often decided by pilot skill and tactics.

Fighters were not a pre-war priority for the US Army Air Force (USAAF) – after all, there was no possibility of air attacks against the American continent.

POLIKARPOV I-16

The Polikarpov I-16 (shown below) served successfully in Spain before the war. It had a good rate of climb and manoeuvrability and was, in some variants, the best-armed fighter anywhere. It made up about half of the Soviet fighter strength in 1941.

SPEED: 460kph (286mph)
RANGE: 440km (275 miles)
ENGINE: Shvetsov M25 radial; 700hp
ARMAMENT: 2 x 20mm (0.79in) cannon; 2 x 7.62mm (0.3in) machine-guns

Right: A cannon-armed Spitfire 5b of 303 Squadron, a Polish-crewed unit, in flight in 1942.

Right: A cannon-armed Spitfire 5b of 303 Squadron, a Polish-crewed unit, in flight in 1942.

The Curtiss P-36 and P-40 had modest capabilities but saw significant service with Britain and France. The P-40 was used extensively by British forces in North Africa in slightly different Kittyhawk and Tomahawk forms, but it could never quite compete with the Bf 109. Later-war USAAF fighters were of much higher quality.

Early Soviet fighters were a mixed bag, made worse by low manufacturing standards and poor pilot training. The mid-1930s' Polikarpov I-16 saw extensive use against Finland and in the early days of Operation Barbarossa. It was reasonably well armed but slow by the standards of 1941. The LaGG-1 and -3 were unusual in being built largely of wood and proved to be rugged but again rather slow in combat service.

The Soviets' MiG design bureau produced the MiG-1 and -3 that were most effective at high altitudes (which must have been an ordeal in the MiG-1's open cockpit), but they were otherwise disappointing. Most important were the various Yak designs. The series reached the Yak-7 variant by 1942 and would see further highly successful development later on in the war.

Japanese Army fighters of the early-war years showed the same characteristics as the better known "Zero" of the Navy (which also served extensively over land). Their excellent manoeuvrability stood them in good stead when faced with the older designs that the Allies

deployed to the Pacific in 1941–2. However, their weaknesses of light construction and inadequate protection for pilot and fuel tanks proved more important against upgraded opposition later. Notable types included the Nakajima Ki-43 "Oscar" and Ki-44 "Tojo".

TWIN-ENGINED TYPES

Perhaps the only twin-engined day fighter to serve successfully was the Lockheed P-38 Lightning, in use from 1941. Its speed and good range meant it

REPORTING NAMES

To avoid confusion resulting from difficulties in pronunciation, American and other Allied forces gave Japanese aircraft "reporting names", which were designed to be short and easily recognizable. Another advantage was that a name could be allocated to a design before its true Japanese designation was known and did not need to be changed thereafter, whatever information later became available. Fighters were given boys' names, while bombers had girls' names.

performed well in the bomber-escort role. Other types like the Bristol Beaufighter or the Messerschmitt 110 lacked the agility that was needed for daytime air combat but appeared in other roles in due course.

DEWOITINE D.520

The Dewoitine D.520 was the best French fighter in 1940. Production failures meant that only 100 were available by May. They fought well in the Battle of France and were later briefly used against the Allies by the Vichy forces in Syria. About 900 were built in all.

SPEED: 535kph (332mph)
RANGE: 1,250km (780 miles)
ENGINE: Hispano-Suiza 12Y45 in-line; 910hp
ARMAMENT: 1 x 20mm (0.79in) cannon; 4 x 7.5mm (0.295in) machine-guns

Air Combat: Weapons, Tactics and Aces

Some pre-war theorists thought air combats were a thing of the past because of the increased speed of aircraft. Instead, sudden dogfights and extended air battles took place in every theatre and air aces became as famous as their WWI predecessors.

Aircraft performance was far from being the only determinant of air combat success during WWII. Fighters and bombers both became more heavily armed as the war proceeded and better tactics were developed and practised.

GUNS AND GUNNERY

Most of the air-to-air weapons in use in 1939 were rifle-calibre (approximately 7.7mm/0.303in) machine-guns. Some fighters, like Italy's CR 42, carried as few as two such weapons, and bombers, like Germany's Heinkel 111, might have only three, in hand-held mounts. Experience soon showed that this was inadequate. The speed

Below: Loading the nose-mounted 0.79in (20mm) cannon in an RAF Bell P-39 Airacobra in 1941.

Above: A hand-trained Vickers K machine-gun, the sole defensive weapon fitted to the Fairey Battle.

of air combats and the fleeting moments when an enemy would be in the gunsight made greater firepower essential.

Britain had realized that something better would be needed to shoot down German bombers. Initially, the Spitfire and Hurricane were fitted with eight machine-guns but this was

also shown to be insufficient because the individual rounds had too little striking power.

Alternatives to the rifle-calibre weapons were heavy machine-guns in the 13mm (0.51in) class and even more powerful but slower-firing 20mm (0.79in) cannon and some still bigger guns. Late-war American fighters generally carried six or eight 0.5in (12.7mm) guns and found this adequate. The Germans, facing numerous Allied heavy bombers, favoured a heavier punch, with weapons fits including 3cm (1.18in) cannon and even 21cm (8.27in) unguided air-to-air rockets.

Late-war bombers like the B-17G Flying Fortress might carry up to 13 x 0.5in (12.7mm) machine-guns in a mix of powered turrets and single mounts, but even that was not enough unless the aircraft also flew in a tight formation with its squadron-mates. The B-29 Superfortress took things a stage further with various of its turrets being remotely controlled, not individually manned.

COMBAT TACTICS

At the start of the war, Britain's Fighter Command instructed its fighter squadrons to use tight formations and planned sequences of manoeuvres to attack enemy bombers. This was soon found to be dangerous and impractical. Shortly, all air forces were using the methods developed by the Germans, in

Right: Ball turret gunners on B-17s (twin 0.5in/12.7mm guns) were all small men for obvious reasons.

particular, in their involvement in the Spanish Civil War – the pair and the finger-four.

The basic unit was made up of a leader and his wingman. They kept relatively close together, with the leader responsible for the tactical decisions and most attacks, while the wingman's principal duty was to make sure they were not surprised from the rear. Two such pairs made up what the RAF called the finger-four, so-called because this group would fly in a loose formation shaped like the spread fingertips of the hand.

AIR ACES

As in WWI, pilots in all countries kept count of their "kills" and successful pilots were celebrated as aces, or in the rather more descriptive German term, *Experten*. Different air forces had varying methods of assessing air combat successes and it is certainly true that pilots generally claimed far more enemy aircraft shot down than were ever actually lost. This was probably as much a product of the speed and confusion of air battle as any deliberate attempt to mislead. However, it is also true that detailed examination of some aces' claims has backed up most of their scores.

By far the highest-scoring pilots were various Germans. The highest-scoring of all was Erich Hartmann with 352, while tens of others claimed more than 100. These high totals reflected the fact that the Germans did not rotate top pilots out of combat to other

duties as often as other air forces. They were also mostly scored in the earlier years on the Eastern Front where enemy aircraft and pilots were relatively poor. One curiosity of the air fighting on the Eastern Front was that it was one of the few combat functions performed by women, though only on the Soviet side – several women became air aces.

Below: David McCampbell was the top US Navy ace of the war. He was eventually credited with 34 victories.

The British Home Front

Although circumstances meant that British troops were never committed to the fight in numbers to match the 1914–18 war, the British people in WWII were more completely mobilized in support of the war effort than in any other country.

Although Britain was and remained a democratic society, governed with the consent of the people, life in Britain during WWII was in many ways as rigorously controlled as in the totalitarian nations. Britain also devoted its national resources more thoroughly to fighting the war than any other country, totalitarian or democratic.

Although in the end Britain was lightly bombed by comparison with the major Axis powers, fears of air attack had important effects from the first. Within weeks of the outbreak of war, over 800,000 unaccompanied children had been evacuated from major cities and billeted with families of strangers in

Above: Since much food was imported, reducing waste was vital.

safer areas of the countryside. Many returned home during the Phoney War only to be evacuated again in the Blitz, and repeat the process during the V-weapon attacks of 1944–5. In all some 6 million civilians, adults and children were involved.

AIR RAID PRECAUTIONS

Various Civil Defence services, including Air Raid Wardens, were organized with hundreds of thousands of members. Public air raid shelters were built or dug in parks; when the sirens sounded Londoners took to the tunnels of the Tube; and 1.5 million householders installed free family refuges called Anderson Shelters. Everyone implemented the blackout at night; it may have done little to misdirect enemy bombers but certainly caused a great increase in road accidents.

Many of the innovations of the war followed from the Emergency Powers Act, which was passed by parliament in May 1940. This act gave the government dictatorial authority over every aspect of life, and it was used extensively. As well as men conscripted into the forces, for example, the government could direct civilian men and women into particular jobs, control their wages in those jobs and forbid them to leave.

Left: Children being evacuated from London in 1939 say their goodbyes to their mothers.

Above: The number of British women working in agriculture more than tripled during the war.

When he took over as Prime Minister in May 1940 and transformed the country's previously lethargic war effort, Churchill formed a coalition government including Labour and Liberal members but principally drawn from his own Conservative Party. Its methods, however, were distinctly socialist in nature, with centralized planning of manpower, food, fuel, the economy and much more. The government was involved in almost every aspect of daily life. The civil service had to double in size to control it all.

Rationing of food, fuel, clothing and other goods was one of the most pervasive effects of the war. Petrol rationing was introduced in the first months of the war, with meat, butter and sugar added in 1940. Staple foods like bread and potatoes were never rationed – and the "Dig for Victory" campaign encouraged people to grow their own food. Factory canteens,

relatively unusual until the war, also supplied many decent meals. Naturally, a black market did develop, but generally speaking the system worked well and was seen to be fair. It was also undoubtedly true that the poorest people, often unemployed and undernourished in the pre-war years, were much better fed during wartime.

THE BEVERIDGE REPORT

Planning for a better future was a major theme even when a successful outcome for the war was still uncertain. The Beveridge Report of December 1942 proposed schemes of "social insurance" to combat unemployment and provide health care. These and other developments would be embodied in what came to be described as the welfare state after WWII.

All in all, people in 1945 felt that their country had behaved well. They had stood up to Hitler from the first and had fought in a united and purposeful manner. They had high hopes for the future, not all of which would be fulfilled.

Churchill (1874–1965) was Britain's Prime Minister from May 1940 until after the end of the war in Europe. He became a government minister in September 1939, after a long period in the political wilderness, because he had been among the first to see the true nature of the menace facing Britain. This and his natural pugnacity won him the premiership the next May. His inspirational leadership would keep Britain fighting over the coming months. At the same time he worked ceaselessly to bring the USA into the war, recognizing far better than anyone else that this alone would bring victory. It is no exaggeration to say that Europe would have entirely fallen to the Nazis without Churchill's courage.

Above: Prime Minister Churchill in a characteristic pose.

Left: Bevin Boys were young men conscripted to work in the coal mines because of the labour shortage later in the war.

Medium Bombers, 1939–41

Whether intended to support ground campaigns or perform longer-range strategic attacks, early-war bombers struggled to achieve a good balance between the conflicting demands of speed, bomb load, range and defensive armament.

There is no exact definition of when an aircraft becomes a bomber rather than a close-support or ground-attack machine, but the "medium bombers" included here are the twin- or three-engined designs used by all air forces of the time for slightly or substantially longer-range missions.

Pre-war air forces almost all believed that formations of bombers each carrying as few as three machine-guns could defend themselves against enemy fighters and go on to bomb their targets accurately by day or by night. Experience would show that these claims were untrue, other than in very exceptional circumstances.

THE BLITZKRIEG ERA

Germany's Luftwaffe seemed to have the most powerful bomber force of the early-war years.

Above: He 111s in the Battle of Britain. Over 7,000 had been built by the time production ended in 1944.

This included three major types. Both the Dornier (Do) 17 and the Heinkel (He) 111 made their combat debuts during the Spanish Civil War, where their speed and the weakness of the opposition made them seem practically invulnerable. This was not confirmed in the Battle of Britain where their weak defensive armament and their modest bomb loads proved more relevant. The third type, the slightly later Junkers 88 (and the upgraded Do 217), were both more capable aircraft.

The contemporary British designs also had their own shortcomings. The Bristol Blenheim Mark 4 had a decent top speed of 428kph (266mph), for example, but only carried 455kg (1,000lb) of bombs. Neither the Blenheim nor its bigger stablemates were well-protected, though some of the larger aircraft included a British innovation of the mid-1930s – the power-operated gun turret. Least satisfactory of all was the Handley Page Hampden, which lost heavily in early daylight

Below: Savoia-Marchetti 79 Sparviero bombers. These were used for torpedo and bomb attacks.

operations and also lacked the range for night strategic bombing. The bigger Armstrong Whitworth Whitley carried a more substantial bomb load (up to 3,175kg/7,000lb) but it was no longer in front-line combat use by late 1943. The Vickers Wellington was much better. Its unusual web-like internal structure gave it enormous strength to go with a reasonable bomb load and speed.

Despite their defeats in 1939–40, various other air forces had aircraft of some potential, though these were seldom available in worthwhile numbers. Poland's PZL P37 was fast and had a good combination of range and bomb load, but the few in service were quickly overwhelmed. France's Farman F223 and Lioré et Olivier 45 both had impressive performance figures and fought well against the odds in 1940.

Various American designs also saw action with British or French forces before December 1941. The most important of these was the Douglas A-20, which was variously known as the Boston and Havoc and was used additionally in significant numbers by the Soviets.

THE MEDITERRANEAN

Italian designs in service in 1940 were essentially the same as those previously used in Abyssinia and Spain. They had been of good quality then, but Italy lacked the resources to develop replacements while also fighting these campaigns. The three-engined Savoia-Marchetti 79 Sparviero and the Fiat BR20 Cicogna were both in this category, though the Cant Z1007 Alcione was better.

DOUGLAS A-20B HAVOC

First used by France in 1940, the Havoc later served with the RAF and USAAF in the attack and night-fighter roles. In all some 9,500 were built up to 1944. The later A-20G variant was very heavily armed.

SPEED: 570kph (355mph)
RANGE: 1,770km (1,100 miles)
CREW: 4
ENGINES: 2 x Wright R-2600 Cyclone radials; 1,600hp each
ARMAMENT: 680kg (1,500lb) bombs; 3 x 0.5in (12.7mm) + 1 x 0.3in (7.62mm) MG

A number of American types also appeared in British service in the Mediterranean, notably the Martin A-22 Maryland and its development, the A-30 Baltimore. Neither saw combat later with the USAAF.

THE FAR EAST

The Japanese Army and Navy both had forces of land-based bombers. Designs that had seen some success in China during the 1930s continued service into the early stages of the Pacific War where they soon proved vulnerable to Allied fighters. These older types included the Mitsubishi Ki-30 "Ann" and the Kawasaki Ki-32 "Mary".

The Mitsubishi Ki-21 "Sally" was a later 1930s' Army design, broadly comparable to Western contemporaries, which included improved defensive armament in later versions. The Navy's Mitsubishi G3M "Nell" and the G4M "Betty" both had long range, but the G4M in particular was poorly protected.

HANDLEY PAGE HAMPDEN

The Hampden came into service in 1938 but was disappointing. Range with maximum bomb load was rather limited and defensive armament was poor. It left the Bomber Command service in 1942. About 700 of the 1,400 built were lost on operations.

SPEED: 410kph (255mph)
RANGE: 1,900km (1,200 miles)
CREW: 4
ENGINES: 2 x Bristol Pegasus XVIII; 980hp each
ARMAMENT: 1,800kg (4,000lb) bombs; 6 x 0.303in (7.7mm) machine-guns

The Battle of the Atlantic, 1939–41

U-boats and Allied anti-submarine forces were in action from the first day of the war to the last. The Battle of the Atlantic was the longest campaign of the war and if Hitler had won it he would almost certainly have won the war, too.

Britain imported roughly half its food, all its oil and many other items essential to the war economy. In addition British forces and the supplies they needed had to be shipped overseas if Britain was to fight its enemies. Since this traffic had to sail to and from British ports, the Atlantic accordingly became a major theatre of war.

THE GERMAN THREAT

Germany's Kriegsmarine was poorly prepared for war in 1939, with few powerful surface ships and a small force of submarines. Only from the spring of 1941

Above: Günther Prien, one of the top U-boat commanders of 1940–1. Prien sank HMS *Royal Oak* in 1939.

would German U-boat strength increase substantially. For the first year or so of the war about a third of German torpedoes failed to explode, which obviously thwarted many attacks.

Drawing on the lessons of WWI, Britain had planned a convoy system to protect its merchantmen, but there were so few escort ships in service initially that this could only be applied to the few hundred kilometres of their journeys nearest the British Isles. Although the escorts did have asdic (later called sonar) for finding a submerged submarine, they did not at first have radar or effective tactics for convoy protection. Britain had few aircraft committed to maritime duties and, until many months into the war, these also did not have radar to find a submarine on the

Raid by *Scharnhorst* & *Gneisenau* Jan–Mar 1941
Raid by *Bismarck* May 1941
Bismarck chase ·········· Aircraft carrier attacks
✕ Major battles

Above: *U-570* in a British port after surrendering to a British aircraft following damage in an air attack.

MAJOR SURFACE ACTIONS OF THE EUROPEAN WAR

The Royal Navy won most battles against the Germans and Italians.

surface nor weapons likely to damage a U-boat that had dived. However, although the German surface and U-boat forces achieved successes up to June 1940, these were not significant enough to pose a real problem.

"THE HAPPY TIME"

The fall of France in 1940 brought a major change. Within hours of the surrender the head of the U-boat force, Admiral Karl Dönitz, had equipment trains rolling to France's Atlantic ports, hundreds of kilometres nearer the convoy routes than previous German bases. What the U-boat crews called the "Happy Time" was about to begin. Until the spring of 1941 a series of U-boat commanders became celebrated as "aces", sinking ship after ship with little loss on the German side.

They used so-called "wolf-pack" tactics whereby the first boat to sight a convoy signalled U-boat headquarters, which then manoeuvred a group into attack positions. Then, at night, the U-boats would sail on the surface right in among the convoy's ships – without radar it was almost impossible to spot a surfaced U-boat. The U-boats would tor-

pedo perhaps several ships and escape into the dark amid the resulting carnage. In this period the Germans also had the upper hand in the code-breaking struggle. Their messages remained secure, but many British ones giving away convoy routes and other movements did not.

Things improved for the Allies in the spring of 1941. Britain began breaking the U-boat codes. Escort ship numbers increased so that convoys could be protected all the way across the Atlantic. The escorts and their few supporting aircraft began to be fitted with effective radar sets. And the USA started taking a more active role, even though still officially neutral.

Above: The German battleship *Bismarck* after sinking HMS *Hood* in May 1941. *Bismarck* is down by the bow because of battle damage.

In April 1941 President Roosevelt extended the Pan-American Neutrality Zone and in July US Marines occupied Iceland. Ships of any nationality travelling between there and the USA were protected by the US Navy. By the autumn the Americans were effectively fighting alongside the British and the much-expanded Royal Canadian Navy. Although Allied shipping losses therefore declined sharply in the second half of 1941, the struggle was clearly by no means over.

SURFACE RAIDERS

The Kriegsmarine also used warships and disguised armed merchant ships in raids on Allied trade routes. Germany's three "pocket battleships" made various generally unsuccessful voyages and one, *Graf Spee*, was sunk in December 1939. The more powerful *Scharnhorst* and *Gneisenau* made a more worthwhile sortie in early 1941. The still bigger *Bismarck* sank HMS *Hood* on 24 May 1941 but was itself hunted down three days later. Seven disguised raiders made a number of voyages during 1940–1 and proved very difficult to track down, operating successfully for extended periods in the Pacific and Indian Oceans as well as the South Atlantic.

Above: A well-ordered convoy is seen from an American patrol aircraft shortly after leaving port.

Anti-submarine Escort Ships

Ships designed for anti-submarine duties did not need to be large or fast or carry numerous guns. Instead the British and US fleets built hundreds of tough and enduring smaller ships, which took a huge toll of enemy submarine forces.

The main anti-submarine vessels in all navies were traditionally destroyers. However, these were expensive to build and their design emphasized speed and anti-ship armament, not endurance and anti-submarine weapons. In World War II, large numbers of smaller and usually slower ships were built, designed and equipped mainly or exclusively for anti-submarine work.

JAPAN'S FAILURE

The major operators of ships in this category were the British and Americans (who supplied ships and designs to other Allied navies and also used each other's designs). The other navy with large ocean-going responsibilities, the Japanese, notably failed in anti-submarine operations, especially in defence of merchant ships. There were two

reasons for this: first, both before and during the war the Japanese concentrated on offensive operations against enemy warships; and, second, they devoted little effort to developing radar and sonar equipment, crucial in the anti-submarine war. Some Japanese escorts had no underwater sensors as late as 1942, a year in which the Japanese Navy ordered the grand total of eight

Above: The destroyer escort USS *Huse* and an escort carrier hunting for Atlantic U-boats in 1944.

escort-type ships. By contrast the Royal Navy built around 600 escorts in its principal classes in the course of the war, while the US Navy produced many more.

Important British escort ship classes were: Black Swan sloops, Hunt escort destroyers, Loch and

---

HMS *LOCH FADA*

Seen here in April 1944 around the time of its commissioning, HMS *Loch Fada* was the first of the 28 Loch-class frigates. The design was a development of the previous River-class frigate, but it had much better anti-submarine weapons and sensors.

COMMISSIONED: 1944
DISPLACEMENT: 1,400 tons
CREW: 114
SPEED: 20 knots
ARMAMENT: 1 x 4in (102mm) + various light anti-aircraft guns; 2 x Squid anti-submarine mortars + depth charge rails and throwers

River frigates and Flower and Castle corvettes. The Hunts and Black Swans were built to standard navy specifications and were effectively smaller and slower destroyers. The Hunt class (86 built) were a little over 1,000 tons standard displacement and could make 27 knots with 4 or 6 x 4in (102mm) guns; some carried 3 torpedo tubes. Fewer Black Swan-class ships were built, similar in size and armament but with a lower top speed (20 knots), in exchange for longer endurance.

The other escort ship classes did not have a purely naval heritage but were designed to be suitable for building in yards without naval experience and to use mercantile engines. In addition the Flower class, the most numerous class of all (267 built), had a hull form based on a civilian whale-catcher design.

The Flower class, with only a single 4in (102mm) gun, did have good anti-submarine weapons and sensors. Disadvantages were that they were slow (at only 16 knots a surfaced U-boat could outrun them) and

were very uncomfortable for the crew in bad weather, common in the winter North Atlantic. The next most numerous type, the River class, were similar in size and performance to the Black Swans but suitable for building in civilian yards. The US Navy's Tacoma class were very similar.

US NAVY TYPES

American production of escort ships was vast. There were six classes of destroyer escorts, over 400 ships built, of similar size and capabilities to the British Hunts. The most numerous of these was the Buckley class, in

Above: The USS *England* (in early 1944) had an amazing success rate.

US Navy service from April 1943. One ship of this type, the USS *England*, achieved the unmatched feat of sinking 6 Japanese submarines within a period of 12 days in May 1944.

Other US escort vessels included a host of smaller types, usually described as submarine chasers. Many of these served with the US Coast Guard and, despite their diminutive size, they too chalked up an impressive record of over 60 submarine kills in the course of the war.

HMS *HONEYSUCKLE*

HMS *Honeysuckle* was a Flower-class corvette built on the Clyde. It is seen here during service with the Arctic convoys in 1945 in the Kola inlet alongside the escort carrier HMS *Trumpeter.* Four Flower-class ships (captured when under construction in France in 1940) were used by the Germans.

COMMISSIONED: 1940
DISPLACEMENT: 925 tons
CREW: 85
SPEED: 16 knots
ARMAMENT: 1 x 4in (102mm) + various light anti-aircraft guns; 2 x depth charge rails (as built)

The Mediterranean Theatre, 1940–1

Thrown out of mainland Europe at Dunkirk, Britain concentrated on the Mediterranean and attacks on Germany's ally Italy in the early-war years. Mussolini's declaration of war in June 1940 would soon seem unwise.

As France was crashing to defeat in June 1940, Mussolini declared war on the Allies, determined not to miss out on a share of the spoils. Italy had annexed Albania without a fight in the spring of 1939 (this had been recognized by Britain as part of the then still current appeasement process). Then, in the summer of 1940, Mussolini picked a quarrel with Greece, which had been trying desperately to stay out of the war. On 28 October Italian troops crossed the border from Albania but their advance into Greece was soon halted and turned back by the Greek forces. By March 1941 half of Albania was under Greek control.

Italy also had large armies in its North African colony of Libya, as well as in East Africa in Italian Somaliland and

Above: Guns and crew of a British A9 cruiser tank in Egypt, May 1940.

Abyssinia. In August 1940 troops from Abyssinia occupied British and French Somaliland. Then, in September, the Italian Tenth Army crossed from Libya into Egypt but it halted and dug in after a short distance.

In all these campaigns the weakness of the Italian forces was apparent. The troops were generally ill-trained and badly

led and had little commitment to the fight. Equipment on land, at sea and in the air had many shortcomings, with flimsy tanks, outmoded biplane aircraft, inaccurate naval guns and more. Results soon made this plain.

ITALIAN DISASTERS

The most spectacular Italian defeat was in the North African desert. By December 1940 the British force in Egypt was ready to respond to the initial Italian advance. Within two months the Italian Tenth Army had been thrown back to El Agheila (Al-'Uqaylah) and lost 130,000 prisoners at a cost of only 550 British dead. However, the

NORTH AFRICA, 1940–2

The Desert War saw a remarkable sequence of changing fortunes, with first one side, then the other on top.

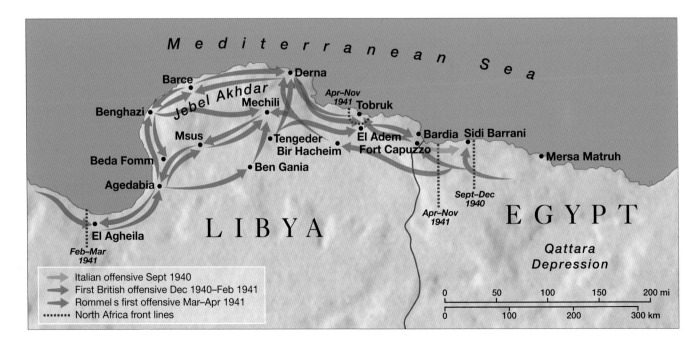

Italian offensive Sept 1940
First British offensive Dec 1940–Feb 1941
Rommel s first offensive Mar–Apr 1941
North Africa front lines

British position was becoming less secure. Troops had already been withdrawn to East Africa; others would soon go to Greece.

ENTER ROMMEL

Embarrassed by his Italian ally's failures, Hitler sent an ambitious young general called Erwin Rommel to Libya in February 1941 with a small force to block any further British advance.

Rommel spotted how weak the British front had become and attacked at the end of March. Within a month the British had been pushed all the way back into Egypt, with their worn-out tanks falling easy victims to the superior German equipment and tactics. Two minor British offensives – at the end of April and again in June – also failed after several further demonstrations of the superior German fighting skills.

Fortunately for the Allied side, one of their campaigns had a much greater success. During January and February 1941 the British, Indian and African troops that had been based in Sudan and Kenya went on the attack into Abyssinia and Italian Somaliland. Most of the Italian troops had been defeated by May though the final surrender did not come until November.

At sea Britain's morale was boosted by substantial gains over the powerful Italian fleet. Early skirmishes went Britain's way. On 11 November 1940, a night air attack by carrier planes on the Italian base at Taranto crippled three battleships.

Right: Some of the tens of thousands of Italian prisoners who were captured by the British in Libya in December 1940.

British forces based on Malta were then able to step up their attacks on the supply routes between Italy and North Africa, contributing greatly to the British success in the desert. In turn, when German air forces arrived in Sicily in strength in early 1941, this helped the land battle swing Rommel's way.

In March 1941 the Battle of Matapan confirmed the weakness of the Italian fleet. Three

Above: Damaged ships in Taranto Harbour, here seen in a British photograph taken after the attack.

Italian heavy cruisers were caught and sunk while on the retreat from Admiral Andrew Cunningham's Mediterranean Fleet. Even though British naval strength declined substantially in the following months, the Italians would not make a major challenge at sea again.

The Balkans and North Africa, 1941–2

In 1941 Hitler established what looked like a secure grip on the Balkan region, protecting his southern flank for the attack on the USSR. Changing fortunes in North Africa brought the British back to a desperate defence of Egypt in mid-1942.

Hitler's main plan for 1941 was to attack the USSR, but he also wanted to secure German control over south-eastern Europe. In the winter of 1940–1 Hungary, Romania and Bulgaria all in effect allied themselves with Germany. Greece, however, was winning its war with Italy in Albania and was receiving increasing help from Britain. This brought the unwelcome prospect of British aircraft based in Greece within relatively easy striking distance of the Romanian oilfields on which Germany greatly relied.

From late 1940, the Germans therefore prepared an attack on Greece. In late March 1941 when it seemed that the Yugoslavian government, after much pressure, was going to join the German bloc, a military coup reversed the situation and a furious Hitler ordered an immediate German attack.

The Germans deployed overwhelming forces for their offensive both in the air and on

Below: A German transport plane crashes over Suda Bay during the invasion of Crete in May 1941.

Above: German troops in northern Yugoslavia during the rapid and successful invasion of April 1941.

the ground. The strike against Yugoslavia started with heavy air raids on Belgrade on 6 April. The main land advance began from Romania on the 8th and was joined in succeeding days by converging attacks from both Hungary and Austria. There was little resistance and the Yugoslavs agreed an armistice on the 17th. German casualties for the campaign were fewer than 200 dead.

By 6 April 1941 three British Empire divisions and supporting forces had been sent to Greece, substantially weakening the British position in North

Africa. However, with much of the Greek Army already committed to the Albanian front and the remainder poorly deployed, there was little chance of the combination resisting the Germans for long. By the end of April the whole of mainland Greece had been overrun. The last act of the campaign was the capture of Crete by German airborne forces on 20–31 May.

DESERT BATTLES

After their defeats in the first half of 1941, it took until November for the British forces in North Africa to prepare a new offensive. They started their advance across the border from Egypt into Libya on the 18th.

A bewildering series of manoeuvres and tank battles followed over the next three weeks or so. Much of the Allied superiority in numbers was frittered away by poor command, but eventually the exhausted German forces and their Italian allies had to retreat. As before this meant retiring all the way to El Agheila on the Gulf of Sirte, which Rommel reached at the end of December.

However, the pendulum was set to swing once more. German air strength in the Mediterranean was increased again and British naval power was at a low ebb, while Rommel's supply situation quickly improved. On land much British strength had been dissipated and resources and troops that might have come

Above: Australian troops like these made up a large part of Tobruk's garrison during the siege in 1941.

ERWIN ROMMEL

Field Marshal Rommel (1891–1944) became famous as Germany's commander in North Africa during 1941–2. Rommel took his first steps to fame when he was given command of a Panzer division in 1940. In 1941 he was appointed to lead German troops in Africa, which he did with considerable skill until outmatched by greater Allied resources in the late summer of 1942. In 1944 he commanded the troops opposing the D-Day landings but was wounded in an Allied air attack shortly after. He was forced to commit suicide later in the year because he had been implicated in the Bomb Plot against Hitler.

Above: Rommel (nearest) and Marshal Italo Gariboldi, the commander of the Italian forces in North Africa in 1941.

to North Africa were instead being sent to face the Japanese advance in the Pacific. Less than a month after ending his retreat, Rommel attacked again.

NEW GERMAN ADVANCE

In the first stage, completed by early February 1942, the British were pushed out of most of the territory they had just captured. Next, in a three-week battle from 26 May, British defences on the so-called Gazala Line were overcome, despite being held by superior forces. Tobruk (Tubruq), which had survived a long siege after Rommel's first advance, was quickly captured this time with a great mass of supplies, which helped support Rommel's *Afrika Korps* for a charge into Egypt. After a disordered and panicky retreat, the British made a stand near an obscure railway halt called El Alamein in the first days of July. Rommel's last desperate attacks were fought to a standstill in the First Battle of El Alamein. Now both sides would settle down to rebuild their worn-out forces.

Torpedo Boats and Midget Submarines

These two varieties of warship were among the fastest and the slowest in service with any navy, but they shared a single quality: they were the smallest vessels capable of sinking an enemy ship of any size.

Victims of midget submarines included several battleships and cruisers in both the Pacific and European wars. Torpedo boats also knocked out a number of cruisers and smaller warships, in addition to having numerous, probably more important, successes against transport vessels of many different kinds in all arenas of war.

TORPEDO BOATS

Varying in length from roughly 24–34m (80–110ft), the torpedo boats in service during WWII generally carried a pair of torpedo launchers and a selection of 20mm (0.79in) or similar guns and lighter weapons, and might reach top speeds of just over 40

knots. Most navies mainly used petrol engines, but the Germans in particular used diesel. The Germans were also unusual in relying on a rounded hull form, whereas most other torpedo boats followed a flat-bottomed style designed for effective planing at high speed.

German torpedo boats were in fact probably the best in service during the war. They were called *Schnellboote* ("fast boats"), but they were usually referred to on the Allied side as E-boats. Their hull shape proved very effective in poorer weather, often a problem for torpedo boats generally. Their diesel engines were also well silenced by having their exhausts underwater and were in any case less prone to catch fire after combat damage than petrol ones. A variety of types

Below: British motor torpedo boats on patrol in the Channel during the Normandy invasion.

USS *PT-174*

PT-174, seen here off Rendova in January 1944, served in the Pacific from mid-1943 to the end of the war. *PT-174* was an 80ft Elco type (326 were built). These wooden boats displaced 56 tons, full load, and carried a variety of weapons in service in addition to the typical list below.

LENGTH: 24.4m (80ft)
SPEED: 41 knots
ENGINES: 3 x Packard 4m-2500; 1,500hp each
ARMAMENT: 1 x 40mm (1.58in) gun, 4 x 0.5in (12.7mm) MG; 4 x 21in (533mm) torpedoes

existed, all rather larger than most Allied designs, armed by the late war with a twin 2cm (0.79in) and a single 3.7cm (1.46in) gun, plus machine-guns and the standard pair of torpedoes.

The American equivalents (also extensively used by the British), the PT for "patrol torpedo" boats, came mainly from the Higgins and Elco companies.

The most common of several slightly different Elco boats were 24.4m (80ft) long, carried four torpedo launchers, a 20mm or 40mm (1.58in) gun and numerous machine-guns. The Higgins boats were slightly shorter and a little slower but were probably more seaworthy. Their greatest successes came not in dramatic attacks on major warships, but in numerous minor operations against Japanese supply barges and similar craft in shallow Pacific island waters where bigger Allied vessels could not go.

MIDGET SUBMARINES

Britain, Germany, Italy and Japan were the main users of midget submarines. Italy and Britain successfully used "human torpedoes" in which frogmen rode on top of a torpedo-like craft and would slowly approach an enemy anchorage to attach explosives to their targets. Britain additionally had X-craft, more like small conventional submarines, carrying massive explosive charges to drop under an anchored enemy ship. The German battleship *Tirpitz* was disabled in one such attack in Altenfjord in 1943.

Other midget submarines relied on firing torpedoes. The smallest were the German *Marder* and *Neger* types, which were in effect manned torpedoes with a second armed one slung underneath. Larger still was the *Seehund* design, a two-man vessel with two underslung torpedoes. These and several other German types appeared in small numbers and achieved scattered successes.

Japan had over 40 midget submarines in 1941, known as the *Type A* or *Ko-hyoteki*. They

Above: A Japanese midget submarine having run aground on a beach in the south-west Pacific.

were approximately 24m (80ft) long and carried two torpedoes. They were used in unsuccessful attacks on Pearl Harbor and on Sydney in Australia. However, they did damage the old British battleship *Ramillies* in a harbour in Madagascar. They were carried close to all these targets by larger "parent" submarines.

SUICIDE WEAPONS

Japan produced numerous suicide surface and submarine craft. Over 6,000 *Shinyo* motorboats were built for use against various US invasion forces. However, these small craft were ineffective – many were hunted down by PT boats. The only submarine weapon to see action was the *Kaiten*, essentially a "Long Lance" torpedo modified to be controlled by a single crewman and designed to be launched from a larger submarine. Several hundred were built but they achieved little.

SEEHUND TYPE XVIIB

The *Seehund* type was the most successful German midget submarine. Some 138 entered service in 1944–45. They probably sank 9 ships, losing 35 of their number. An advantage of their small size was that when depth-charged, they might be thrown aside rather than being damaged.

DISPLACEMENT: 17 tons
CREW: 2
SPEED: 7 knots
ENGINES: 60hp diesel; 25hp electric motor
ARMAMENT: 2 x 533mm (21in) torpedoes (carried externally)

The USA and the European War

For over two years after Hitler attacked Poland the USA remained neutral, though all the time it sent increasing help to Britain and the other Allies. Even so it took Hitler's declaration of war to confirm that the USA would fight Germany.

Since the end of WWI the United States had returned to its traditional foreign policy of keeping out of overseas conflicts and maintaining only modest armed forces. By September 1939 this "isolationism" had been backed by strict neutrality laws that prevented the US government or private corporations from selling arms or giving loans to countries at war. Most Americans blamed Germany for starting the war and hoped that Britain and France would win it, but they were also very clear that the USA should stay out.

President Roosevelt and his government saw things a little differently. They recognized the evil of the Nazi regime, and of Japan's militarism, and that (as well as being morally repugnant)

Below: In September 1940 in the "destroyers for bases" deal, Britain was given old US Navy ships like these to fight the U-boat menace.

Above: Pioneer aviator Charles Lindbergh, one of the leaders of the isolationist America First movement.

they presented a real threat to the USA's interests. Roosevelt no more wanted war than did the American people so his policy was to help the European Allies with "all means short of war", while being as tough with the Japanese as possible without provoking them to attack. The

details of this changed over time but the general principles held good up to December 1941.

NOT QUITE NEUTRAL

The first step came in November 1939 when the "cash and carry" law was passed. Countries at war could buy American arms with cash (not on credit), provided they transported them overseas on their own ships. The facts of geography and Britain's large navy meant that this would only benefit the Allies, as was intended.

The hope that this would be enough to ensure an Anglo-French victory fell apart with France's surrender in June 1940. The USA's rearmament was immediately stepped up, in particular by the Two-Ocean Navy

Below: Roosevelt took leading Republicans into his cabinet, including Frank Knox, here being sworn in as Navy Secretary.

FRANKLIN D. ROOSEVELT

Roosevelt (1882–1945) was President of the USA from 1933 to his death. While Churchill's wartime leadership is almost universally admired, Roosevelt's has always been more controversial. Although he was no warmonger, he saw clearly the threats posed by Germany and Japan and worked steadily to oppose them. Opponents, however, said that he deceived the American people over just how far he was going. He has also been criticized as being naive in his dealings with Stalin and the Chinese Nationalists. On the other hand his generous instincts were shown in his Lend-Lease policy, without which Britain and other Allies could not have continued the fight.

Right: Roosevelt giving one of his famed "fireside chat" radio broadcasts.

Act of July, which provided for a massive increase in American naval forces. This was designed to protect against the German threat but it also worried Japan. It would take a couple of years to build the ships but by then the Japanese would be unable to compete, unless they did something about it first.

Next, in the autumn of 1940, came the Selective Service Act, establishing conscription for the first time in US history when the nation was not at war.

LEND-LEASE

Roosevelt also continued to help Britain, although this carried risks – sending arms to Britain might mean Hitler getting them for free if he won the war. However, there was a new problem – Britain was about to run out of cash and would probably be unable to continue the war at all, never mind pay for more American goods. The answer was the "Lend-Lease" programme, in operation from March 1941. Britain, and later other Allies including the USSR, would be sent vast supplies of arms and other goods, produced

initially at American expense, on the basis that they would be paid for or returned after the war.

Roosevelt was well aware that there was no point making arms for Britain only for the Germans to sink them on their way across the Atlantic. Through 1941, therefore, the US Navy played a more active part against Germany's U-boats in the Atlantic, though how far this was going was not made clear to the American people. By autumn US warships in the western Atlantic were doing the same things as the British and Canadians. It was equally clear,

though, that Americans still wanted to avoid war. In August 1941 the Selective Service Act was only renewed in Congress by a single vote.

Throughout all this Hitler had been cautious, content to fight only the enemies he already had. Incomprehensibly, four days after Pearl Harbor, he changed his mind and declared war on the USA, a decision that doomed his regime to a defeat as crushing as that awaiting Japan.

Below: Men register for the draft in October 1940, under the terms of the new Selective Service Act.

German Rule in Europe

Life in the countries occupied by the Nazis was exceedingly harsh. Starvation rations, slave labour and other cruelties were everyday realities in all the occupied territories, though Jews and resisters fared worse still.

Racism was the whole basis of Germany's rule in Europe. Everything was to be done and organized for the benefit of both Germany and the *Volk* – the German race, as defined by Hitler and the Nazis. Within this system there were gradations. At the top were those like the Dutch or Norwegians, who were certainly regarded as second-class citizens but still worthy of some respect, and near the bottom were the Slavs, whose lives were valueless even if sometimes their labour was not. Jews were the lowest category of all.

The way in which conquered territories were ruled naturally varied within this hierarchy. In Poland or the Ukraine Nazi cruelty, organized and led by the SS, was open and extreme.

Above: A German officer and a British bobby in the Channel Islands, the only occupied British territory.

Above: *Volksdeutsche* from Bessarabia in Romania, on their way to be resettled in Germany.

Near the other end of the scale were countries like Norway, where local Nazi sympathizers were allowed the appearance of a say in the government. (Norway's Vidkun Quisling has given the English language a word for just this sort of traitor and was executed for his treason after the war.) Finally, there were nations like France where a government at least not hostile to Germany was allowed to control all or part of the country.

The "independence" of Marshal Pétain's Vichy regime was bought at a price, however. More than half of government revenue went straight to Germany to pay for the costs of the occupation forces and almost half of French industrial output was for German benefit. The franc was artificially valued,

so that anything the Germans did pay for was acquired on unfair terms.

Food rationing was severe with the official provision set at an inadequate 1,200 calories per day, though many of the poorest did not even manage this – much food was diverted to an expensive black market. The experience of other Western European countries was similar in many of these respects.

FORCED LABOUR

As the war proceeded Germany became increasingly dependent on foreign labour to keep its economy running. This included around 1.5 million prisoners of war and, by 1943–4, some 5 million civilians, most of whom had been forced to work in Germany and many of whom were treated as slaves. Roughly a quarter of the workforce in Germany in the later war years was made up of foreigners.

The process started in 1939–40 when many Poles were brought to Germany as farm labourers. Later, in eastern Europe, men, women and many young children were simply rounded up and sent wherever it suited Germany. In Vichy France men were conscripted to do their national service in the German labour force.

VOLKSDEUTSCHE

People of German descent (known to the Nazis as *Volksdeutsche*) were living in many

Above: German officers enjoy a Paris café in the gentler early days of the occupation in 1940.

places in eastern Europe; one of the most bizarre aspects of Nazi rule was the plan to reassimilate them with the *Volk*. Hundreds of thousands of people were to be brought back into the Reich. Many were transported from their homes in the Baltic States or western USSR, supposedly to be resettled in captured territories that were to become part of Germany proper. Most ended up among the millions of homeless displaced persons in central Europe after the war.

Perhaps the most heartless aspect of this particular policy was the *Lebensborn* programme. As part of this plan to increase the German race, SS representatives toured occupied territories identifying "racially worthwhile" children and taking them for forced adoption in Germany. Some 300,000 children are believed to have been abducted from their families in this way; 80 per cent of them never returned.

Above: Reichsführer-SS Heinrich Himmler on parade.

Left: Hitler with his fervent admirer Vidkun Quisling. The Germans never gave Quisling any real power in occupied Norway.

Operation Barbarossa

Germany's invasion of the Soviet Union was the greatest military operation ever seen. Its outcome would decide the result of the war and the campaign would be fought on both sides with a vicious barbarity unequalled elsewhere.

Hitler had first ordered his generals to begin planning an attack on the Soviet Union even before the Battle of Britain was fought. Ultimately, the Nazis intended to occupy the whole of the European USSR, exterminating all Jews and communists and enslaving any surviving "sub-human" Slavs.

Although the Germans deployed a truly massive force of around 3 million men, 3,300 tanks and 2,700 aircraft, the task facing them was huge. They were outnumbered by the Red Army forces in the western USSR, and they greatly underestimated the size of the Soviet reserves in the Far East and elsewhere. They thought little of Soviet weapons or efficiency; indeed, only a quarter of the total 24,000-strong Soviet tank force was in running order – and German intelligence knew little of the superior Soviet weapons like the KV-1 and T-34 tanks, which were coming into service.

GERMANY'S PROBLEMS

The huge distances were daunting. The Germans had pressed hundreds of vehicles from every conquered nation into service but still needed over 600,000 horses for their transport needs. Most ordinary soldiers simply had to march. Roads everywhere were unmade or in dreadful condition, and the Soviet rail system ran on a different gauge, so the track would need to be converted before it could be used. The

German movements
Front line 22 June 1941
Front line 16 July 1941
Front line 25 Aug 1941
Front line 30 Sept 1941
Front line 5 Dec 1941
Trapped Soviet troops

Below: Soviet troops fought with undoubted bravery in what they soon called the Great Patriotic War.

OPERATION BARBAROSSA
Neither Hitler nor his generals could decide which of their three main advances should have priority.

Above: Muddy roads across the endless steppe, with Panzers at war in the east in 1941.

General Guderian (1888–1954) was the man most responsible for the creation of the German Army's formidable Panzer arm. His pre-war book *Achtung! Panzer!* explained how tanks might be used in a new fast-moving style of war. He was highly successful in command of a Panzer corps in both Poland and France in 1939–40. These successes were repeated during the early months of Operation Barbarossa, but he was dismissed by Hitler that winter. He served in staff positions later in the war, but by then Hitler was making most of the decisions and Guderian's advice was seldom followed.

problems in these areas would escalate the farther the advance went and as the bad weather of autumn and winter set in.

The Germans' greatest asset was that, despite many warnings, Stalin refused to allow proper preparations to be made and no one dared to argue. Why this was is a mystery. He seems to have been so frightened by the chance of a German attack that he tried everything he could to avoid provoking one.

THE BATTLE BEGINS

The German attack, Operation Barbarossa, began early on the morning of 22 June 1941. Within hours the leading tank forces had penetrated tens of kilometres into the unprepared Soviet defences and over a thousand Soviet aircraft were wrecked on their airfields. Within a week the two Panzer groups leading Army Group Centre's advance had completed a vast encirclement west of Minsk; 300,000 Red Army troops surrendered there by early July. In the meantime the attack had surged on toward Smolensk, another 320km (200 miles) nearer Moscow.

Progress on the other main attack front was also excellent; Army Group North had over-run the Baltic States and was closing in on Leningrad. The less powerful Army Group South, however, was being held up by Soviet resistance just west of Kiev. Neither Hitler nor his generals had been sure about the best strategy for the campaign from the start and now Hitler ordered a change of plan. Guderian's Second Panzer Group was to attack south, to help complete the capture of Kiev and not, as Guderian and other generals urged, immediately continue east to Moscow.

The results seemed to justify the decision. Over half a million Red Army troops were captured east of Kiev by mid-September. The ever wearier German troops redeployed to the Moscow front by the end of the month for what they hoped would be the final decisive attack.

However, the delay had brought the start of bad weather. After yet more fierce fighting, they reached Moscow's outer suburbs in early December but they could go no further. The German units were now a fraction of their former strength and the troops were suffering

Above: General Guderian with his *Panzergruppe* in 1940.

horribly from the freezing weather. A quick victory had once seemed so certain that no supplies of winter clothing had been prepared. Instead a long, horrible struggle on the Eastern Front was now inevitable.

Anti-tank Guns, 1939–42

As tanks were steadily upgraded so, too, were the anti-tank guns used against them by the infantry and artillery. The first anti-tank guns were small, manoeuvrable and easy to conceal, but as they grew in size these qualities came under threat.

When tanks were initially introduced in WWI, the weapons used to counter them were either standard artillery guns or specially powerful rifles. Although anti-tank (AT) rifles were still used by most armies in 1939, it was clear in the 1930s that more powerful specialized AT weapons were needed by the infantry and other forces.

Germany's standard AT gun at the start of the war was a 3.7cm (1.46in) weapon made by the Rheinmetall company from 1936. This gun (formally the Panzerabwehrkanone [PaK] 36) was typical of those in service with other armies, too. It was mounted on a wheeled carriage, giving an overall weight of under half a tonne, and could readily be manhandled in action. It fired a 0.68kg (1.5lb) armour-piercing (AP) shot, cap-

Below: A Soviet 45mm (1.77in) M1937 anti-tank gun under fire.

able of defeating 31mm (1.22in) of armour angled at 30 degrees at a distance of 500m (550yd).

SOVIET AND US TYPES

The Soviet Model 1930 37mm gun had originally been developed by Rheinmetall and was very similar to the PaK 36. Japan also had a licence-built version of the PaK 36. The US Army M3A1 37mm, introduced in 1940, was slightly different, though examples of the German gun were studied by the American designers. Over 18,000 of these were produced. Although it was outclassed in Europe by late 1942, this gun saw most service in the Pacific against Japan's weaker tanks.

In 1939 Britain's standard weapon was the 2pdr (40mm), with similar performance, though on a rather heavier and more elaborate carriage. The Soviets also had the 45mm (1.77in) M1937, which was

slightly more powerful – and it was later replaced by a longer-barrelled M1942 version.

HEAVIER CALIBRES

By 1939 heavier weapons were being developed in the West. Germany stepped up to 5cm (1.97in) in the PaK 38, with more than double the armour penetration of the PaK 36.

3.7CM PAK 36

The 3.7cm PaK 36 anti-tank gun (shown below during a river-crossing operation in the Netherlands in 1940) was the German Army's standard weapon at the start of the war. It was replaced in most units by 1942. A similar weapon was mounted on the Panzer 3 and various other armoured vehicles.

CALIBRE: 3.7cm (1.46in) L/45
MUZZLE VELOCITY: 762m/sec (2,500ft/sec)
ARMOUR PENETRATION: 31mm (1.22in) at 30° at 500m (550yd)
WEIGHT OF SHOT: 0.68kg (1.5lb)

Britain's next type was the 6pdr (57mm), but this was slow to come into service because the switch in production from the 2pdr was delayed by the need to re-equip the army after Dunkirk. The USA's 57mm M1 was essentially a 6pdr manufactured under reverse Lend-Lease. There was also a Soviet 57mm M1943 weapon, produced in relatively limited numbers by Soviet standards. All these nations would make still bigger guns later in the war.

Germany was unusual in the early-war period in having both smaller- and larger-calibre

Above: British troops training with a 2pdr (40mm) anti-tank gun in 1942.

8.8CM FLAK 36

Throughout the war Allied tank crews feared the famous "eighty-eight" above all other German weapons. The example shown below is seen in service at El Alamein in 1942. The open spaces of the Desert War favoured the accuracy, armour penetration and range of this powerful gun.

CALIBRE: 8.8cm (3.46in) L/56
MUZZLE VELOCITY: 773m/sec
(2,536ft/sec)
ARMOUR PENETRATION: 110mm
(4.33in) at 500m (550yd)
at 30° (99mm at 1,000m)
WEIGHT OF SHOT: 10.2kg (22.5lb)

weapons in regular AT use. The larger weapons were 8.8cm (3.46in) guns, originally produced for anti-aircraft (AA) service as the FlaK 18 and 36. As AA guns these already had the high muzzle velocity that was needed for the AT role and, unlike the AA weapons in most other armies, they were supplied with appropriate ammunition and specifically designated for such duty. These were by far the most formidable AT guns of the early-war period.

The smaller-calibre weapon was a so-called "squeeze-bore" design, in which the barrel tapered in size from breech to muzzle with a specially designed round being compressed as it passed down the barrel; the resulting build-up in pressure produced a very high muzzle velocity.

The only weapon of this type that saw significant service was Germany's sPzB 41, 2.8cm tapering to 2cm (1.1–0.79in). It had a similar AP performance to the PaK 36 and in the version for airborne troops weighed only 118kg (260lb). Its disadvantage was that it required tungsten-cored ammunition. Tungsten was very scarce in blockaded Germany, so production of these guns was ended in 1943.

This, however, was a pointer to future developments. Most early-war AT rounds were relatively simple solid-shot designs. From around 1942 capped designs and composite construction increasingly took over and would be extensively used in late-war weapons.

Horrors of War: The Eastern Front

The war on the Eastern Front was conducted with a savageness hardly seen elsewhere during WWII. The Nazis saw their enemies as valueless racial inferiors and to the Soviet leaders life was cheap – their own citizens' or anyone else's.

As the German drive on Moscow ground to a halt in temperatures of minus 40 or worse in early December 1941, the pitiless pattern of the war on the Eastern Front was already all too well established. Millions of civilians and prisoners of war (POW) would die near the front and behind the lines as a consequence of the fighting.

MASS MURDER

Before Operation Barbarossa began Hitler had ordered that any captured communist officials were to be killed. As it had done during the invasion of Poland, the SS formed *Einsatzgruppen* to follow behind the advancing armies. These murder squads (about 3,000 strong) had orders to eliminate all Jews and communists. By the end of 1941, as their own

Above: A Red Army anti-tank gun crew advancing any way they can during the Battle for Moscow.

records attest, these squads had killed over 600,000 Jews in the occupied USSR.

POW on both sides also faced a grim fate. Over 5 million Red Army soldiers were taken prisoner by the Germans during the war, 3.8 million of them in 1941; only around 35 per cent of these survived their captivity. Prisoners on both sides were given little food. The only medical attention they received was from their own captured medical staff and any supplies they had.

It is likely that some of this maltreatment arose from lack of resources and incompetence, but the racism at the heart of Nazi ideology and the total disregard for human life central to Stalinism played a larger part. And, of course, many more who tried to surrender were simply shot out of hand. The Waffen-SS (a substantial portion of the German forces later in the war)

almost never took prisoners on the Eastern Front. The Soviet attitude to POW and potential enemies was clearly shown by their murder at Katyń and elsewhere of some 15,000 Polish Army officers who fell into their hands in 1939.

Around a fifth of the population of the German-occupied territories either fled as refugees or were evacuated by the Soviet authorities. Many were sent to work in the new factories being set up east of the Urals, where living and working conditions were harsh in the extreme. Significant numbers of those who remained behind might well have welcomed the Germans had they been well treated. Instead their lives and property were always at risk –

Below: A German soldier gives himself up, winter 1941–2. Many German POWs died of ill treatment.

Right: An exhausted German soldier asleep in his muddy trench on the Eastern Front in early 1942.

whole villages were routinely burnt to the ground in retaliation for partisan activities. On the other hand partisans were often unpopular with civilians; they were tightly controlled by the communist hierarchy and quick to "discipline" those who failed to follow their lead.

MOSCOW ATTACK

Some of the most pitiless fighting of the campaign took place in the winter of 1941–2. On 6 December the Soviet forces began a successful counter-attack on the Moscow front, using fresh reserve forces assembled from the Far East. Unlike the Germans, the Soviet troops had good winter clothing and equipment fit to withstand the country's extreme cold.

Hitler's response was to order "fanatical resistance" and no retreat. It is possible that his order to stand fast was correct in this case and saved his forces from disintegration, but various

of his best generals disagreed and were dismissed. Hitler took over himself as Commander-in-Chief of the Army and would now personally control all major military decisions on the Eastern Front. Although in places they were pushed back

some 160km (100 miles), early 1942 saw the Germans solidifying their front line until all operations were halted as usual in the east by the spring thaw.

Below: Mass graves of some of the Polish officers murdered at Katyń.

Below: German soldiers beating a Russian villager as others watch.

Battleships and Battle-cruisers The USS *Indiana* shelling the Japanese coast in 1945.

Coral Sea and Midway Aircraft aboard the carrier USS *Enterprise* during the Battle of Midway.

Japan's Continuing Successes Burmese civilians look on as Japanese invaders pass by.

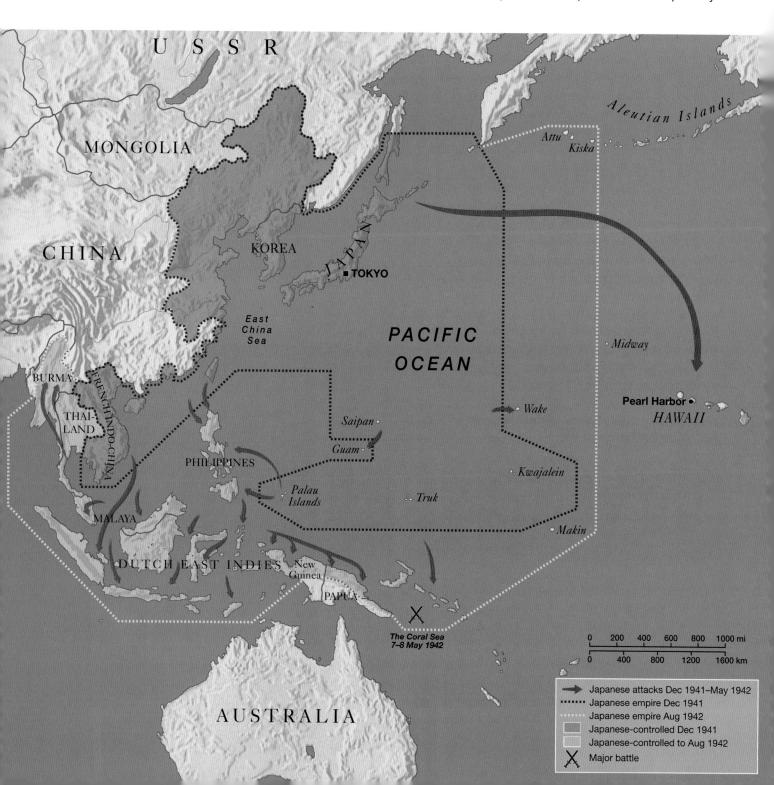

USSR

MONGOLIA

CHINA

KOREA

JAPAN

TOKYO

East China Sea

PACIFIC OCEAN

Aleutian Islands

Attu
Kiska

Midway

Wake

Pearl Harbor
HAWAII

BURMA

FRENCH INDO-CHINA

THAI-LAND

Saipan

Guam

PHILIPPINES

Palau Islands

Truk

Kwajalein

Makin

MALAYA

DUTCH EAST INDIES

New Guinea

PAPUA

The Coral Sea
7–8 May 1942

AUSTRALIA

| 0 | 200 | 400 | 600 | 800 | 1000 mi |
| 0 | 400 | 800 | 1200 | 1600 km |

→ Japanese attacks Dec 1941–May 1942
······ Japanese empire Dec 1941
······ Japanese empire Aug 1942
     Japanese-controlled Dec 1941
     Japanese-controlled to Aug 1942
X Major battle

JAPAN'S PACIFIC CONQUESTS

Japan's attack on Pearl Harbor began a stunning half year of military triumphs that saw the overrunning of many British, American, French and Dutch colonies – from Burma in the west to a range of Pacific islands in the east and south. As well as the victories on land and the decimation of the US Pacific Fleet, British and other naval forces were practically wiped out throughout the theatre of operations and Allied air power was largely neutralized.

Japanese leaders had hoped that the conquest of these territories would give access to the natural resources their country desperately wanted and create a defensive perimeter that could be defended successfully against the worst their enemies might do. Instead, just as in their earlier and ongoing involvement in China, they were sucked into yet more advances, which confirmed what the most sensible Japanese commanders had long realized: in attacking the USA, as previously in China, they had bitten off far more than they could chew.

Japan also made much of plans for a Greater East Asia Co-Prosperity Sphere, a vague anti-imperialist alliance from which all native Asians might hope to benefit. In fact the expanded Japanese Empire created in 1941–2 turned out to be brutal, racist and exploitative.

The USA at War Making aircraft engines was a new area of work for American women.

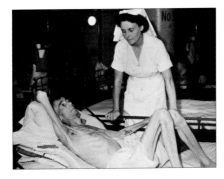

Japan's Brutal Empire A sick British ex-prisoner gets medical treatment after his liberation in 1945.

Japan's First Advances General Percival signs the surrender of Singapore on 15 February 1942.

Approach to War in the Pacific

Throughout 1939–41 tension between Japan and the United States grew. Its armies increasingly bogged down in China, Japan decided to look south for the natural resources and the empire its leaders thought their country needed and deserved.

In early 1939 Japan's leaders were unsure what strategy to adopt to overcome their difficulties in China and their related economic problems. Some wished to attack north against the USSR, others to move south towards the European colonies of South-east Asia, with their substantial natural resources of oil, tin, rubber and much more.

VITAL SOVIET VICTORY

Japan had a long-standing enmity with Russia. Through the 1930s there had been frequent battles along the frontier between Japanese-controlled territory in north-east China and the Far Eastern provinces of the USSR. In the summer of 1939 there was particularly fierce fighting in the Khalkhin Gol region on the border between Soviet-backed Mongolia and China – and the Japanese were badly defeated.

These little-known battles were a decisive turning point in history. Both the Japanese Army and the government began looking more urgently at expanding to the south. In April 1941 Japan negotiated a Neutrality Agreement with the

Left: US Secretary of State Cordell Hull with Japan's Ambassador Nomura during the final pre-war negotiations in 1941.

Soviets and would stick to this decision, even when tempted to change it by Hitler's attack on the USSR in June 1941.

Tension between Japan and the USA and European colonial powers over events in China had grown in the 1930s. Japan was convinced that supplies reaching China through Burma and French Indo-China were vital in keeping the Chinese fighting. Then, in the summer of 1940, new possibilities opened up. Hitler's victories left the French and Dutch colonies virtually defenceless and British power in Asia gravely weakened while the USA's rearmament still had a long way to go.

In July 1940 the Japanese government formally decided on a twin policy: first, to win their existing war by blocking the supply routes to the Chinese; and second, to gain access to the desired raw materials from Malaysia and the East Indies, if necessary by a new war. This was the moment when war between Japan and the USA became more likely than not. In September Japan took the first step in this plan and moved troops into northern Indo-China, by agreement with

Left: Germany, Japan and Italy sign the Tripartite Pact, September 1940.

the Vichy French. In return the Americans imposed an embargo on some iron and steel exports to Japan. A retaliatory process of escalation had begun.

Diplomatic preparations for new advances came with Japan's signing of the Tripartite Pact with Italy and Germany in September 1940. This was clearly aimed, from Japan's point of view, at limiting the USA's responses in the Pacific by the threat that such responses might involve the USA in a European war.

INTO INDO-CHINA

In July 1941 a Japanese government meeting decided to go into southern Indo-China as a first step to further moves south, even if this meant war. However, in the meantime, Japan would continue negotiations with the USA.

American code breakers read reports of the meeting; US leaders concluded that Japan had passed the point of no return, though in fact the Japanese were still uncertain. The Americans decided to step up their economic pressure, cutting

off almost all of Japan's oil supplies. This convinced the more militant Japanese that the USA had aggressive intentions, while at the same time they saw their stockpiles of strategic materials gradually diminishing.

Half-hearted negotiations continued into November but by then the new Japanese Prime Minister, General Tojo Hideki, and his government were convinced that their country had dithered for long enough. There seemed no serious possibility of agreement with the Americans without an unthinkable Japanese withdrawal from China. Japan still had a fighting chance, or so they believed, but they must take it immediately. On 29 November they made the decision to go to war.

Left: Soviet troops celebrate their decisive victory over Japan at Khalkhin Gol in 1939.

THE JAPANESE GOVERNMENT

To outsiders, Japan looked to be run in a similar way to countries like Britain, with an elected government and a ceremonial monarch. In fact in Japan the Army and Navy, and even factions within them, largely controlled national policy. The most militant officers were often those with least knowledge of the wider world, but their supposed patriotic motives were widely respected. Men of this kind had provoked the "Manchuria Incident" as far back as 1931, and ten years later their underestimation of US power would lead Japan into another disastrous war.

Above: Prince Konoye and General Tojo, Japan's last premiers before December 1941.

Left: Japanese troops extending their occupation of French Indo-China in August 1941.

Pearl Harbor and Japan's War Plans

Japan seemed to win a stunning victory at Pearl Harbor, but in the longer term the attack on sovereign United States territory, without a declaration of war, could hardly have been less in Japan's interests. The USA would fight on until victory.

On 7 December 1941 planes from six Japanese aircraft carriers made a surprise attack on the main base of the US Pacific Fleet at Pearl Harbor in Hawaii. Japan had not delivered a declaration of war when the attack began; the USA declared war on Japan the next day.

WHY PEARL HARBOR?

Japan's decision to attack Pearl Harbor came about in a round-about way. What Japan wanted was control of the natural resources of both the British and Dutch colonies in Malaysia and the East Indies.

KEY FACTS

PLACE: Oahu Island, Hawaii

DATE: 7 December 1941

OUTCOME: US Pacific Fleet crippled; USA and Japan go to war.

Below: Pearl Harbor at peace shortly before the Japanese attack. The Ford Island airstrip (centre) and "Battleship Row" to its left would be prime targets, but the oil storage tanks, further left, would be missed.

The Philippines, an American commonwealth, stood on the flank of such an advance and therefore would be attacked too, so war with the USA certainly had to be anticipated.

With Britain busy with the European war and the other colonial powers' homelands under German occupation, the main threat to Japan would come from the US Pacific Fleet – the US Navy was formidable though the US Army was not. Hence a strike on Pearl Harbor at the outset would gain time for Japan to grab the territory it sought. Japan would then fortify

Above: Japanese aircrew board their planes, ready to attack Pearl Harbor, 7 December 1941.

Above: The Ford Island airfield after the attack as ships in the harbour burn in the background.

a perimeter round its new empire which, Japan's leaders believed, would be so daunting that the feeble and decadent Americans would not dare to attack but instead make peace.

AMERICAN PRECAUTIONS

In the weeks before the attack the US government knew from code-breaking information that a Japanese military attack was very likely imminent. They thought that the most likely target would be the Philippines, and that the main danger to Hawaii was sabotage by Japanese agents. However, since the war it has been alleged that President Roosevelt and his government knew of the Japanese plans but did not take proper precautions, so they could mislead the American people into going to war. No real evidence for this claim exists.

From about 07:45 hours that Sunday morning, two waves of Japanese aircraft struck the naval anchorage and various air-fields on Oahu for two hours. By the end of the strikes, 2,403 Americans were dead, 6 of the 8 US battleships in port had been sunk, along with other vessels,

YAMAMOTO ISOROKU

Admiral Yamamoto (1884–1943) was Japan's principal naval commander of the war until his death in 1943, when his aircraft was intercepted by American fighters acting on code-breaking information. He had played a large part in the development of Japanese naval aviation before the war and was the architect of the Pearl Harbor attack. However, he had opposed going to war with the USA, saying that his fleet would win victories for the first few months, but after that Japan would certainly be crushed. His predictions were absolutely correct.

Above: Admiral Yamamoto working at his chart table.

and 188 American aircraft were destroyed. The Japanese lost 29 aircraft, 5 midget submarines and 1 larger submarine. On the US side a catalogue of errors had made Japan's task easier: radar warnings were ignored; anti-aircraft ammunition boxes were locked; and aircraft were easy targets on the ground because they had been parked together so they could be easily guarded against sabotage.

For the future there was now no doubt that, outraged by what President Roosevelt called "a day of infamy", the American people would fight and would continue their fight until their overwhelming resources had brought a total victory. And they also still had significant means to hand. The Pacific Fleet's air-craft carriers happened to be away from port that Sunday, a chance survival that would help point the way to the methods by which Japan would be defeated.

It was not just the Japanese who acted against their own best interests at Pearl Harbor. On 11 December 1941 Hitler declared war on the Americans. The USA now had to fight in the European war as well.

Battleships and Battle-cruisers

These "capital ships", with the biggest guns and thickest armour, were traditionally seen as the decisive weapons in naval warfare. WWII saw few battleship versus battleship engagements, but these vessels still played a vital part.

The navies of the major countries fought WWII using a mix of battleships built during WWI and the 1920s and more modern vessels from the late 1930s onward.

In the inter-war period the number of battleships in service and the maximum size of new ships was limited by the Washington Naval Treaty of 1922 and other later agreements. Britain, the USA and Japan were permitted 15, 15 and 10 ships, respectively, with lower numbers for both France and Italy. Germany would also join the naval arms race, which developed as war approached, but no other navy had battleships of any significance.

Below: The USS *North Carolina* off the American coast in 1942, shortly before joining the US Pacific Fleet.

The ships surviving from the WWI era were of two types: battleships (retained by all major navies) and battle-cruisers (Britain and Japan only). Both had similar armaments and were broadly similar in size, but battle-cruisers had thinner armour in a trade-off for higher speed, a combination that had not always proved successful in WWI. Ships commissioned during WWII combined high speed and thick armour, and accordingly increased in size.

The USS *Mississippi* (New Mexico class, completed in 1917) was typical of the "WWI ships" still in service in 1939. On a displacement of 33,000 tons it carried a main armament of 12 x 14in (355mm) guns, could steam at 21 knots and was protected by side armour up to 355mm (14in) thick.

All older capital ships serving in 1939 had been modernized in the inter-war years to some degree. These changes were made principally to improve their anti-aircraft (AA) capabilities. More anti-aircraft guns were fitted, some of them new dual-purpose (DP) designs suitable for use also against surface vessels. For example, HMS *Queen Elizabeth* began the war with 20 x 4.5in (114.3mm) DP

HMS *DUKE OF YORK*

Third of the five King George V-class ships, *Duke of York*'s most notable achievement was the sinking of Germany's *Scharnhorst* in late 1943. The class's unusual main gun set-up of two quadruple turrets and a twin was troublesome at first but eventually efficient.

COMMISSIONED: 1941
DISPLACEMENT: 37,500 tons
SPEED: 30 knots
BELT ARMOUR: 15.4in (39cm) max.
ARMAMENT: 10 x 14in (355mm) + 16 x 5.24in (133mm) guns

secondary guns, compared to the 2 x 3in (76.2mm) AA guns installed when the ship entered service in 1915. Deck armour was also commonly increased to improve protection against bombs and long-range shellfire. These changes tended to increase displacement (forbidden by treaty), which was partly compensated by fitting lighter but more powerful engines.

LAST TREATY SHIPS

As war approached both Britain and the USA began building modern battleships that reflected these trends. They also attempted to keep within the treaty limitations.

The US Navy's two North Carolina and four South Dakota-class ships carried 9 x 16in (406mm) and 20 x 5in (127mm) DP guns on a displacement (as built) of 37,000 tons; the top speed was 28 knots. Britain's King George V class (the final British class to see war service) was broadly comparable, with slightly less main gun power and slightly thicker armour. America's late-war Iowa class (four ships) displaced a third more than the earlier designs and had a very high speed to escort the aircraft-carrier task forces, which by then formed the heart of the US fleet.

AXIS GIANTS

Germany's *Bismarck* and *Tirpitz*, though begun before the war, made no attempt to keep within the treaty limitations. Both ships were over 42,000 tons and were very well-protected, but

Right: The USS *Indiana* (South Dakota class) shelling the Japanese coast in 1945.

TIRPITZ

Like its sister ship *Bismarck*, *Tirpitz* achieved little for Germany. *Tirpitz* never fired its main guns against an enemy ship and was finally sunk by the RAF in 1944.

COMMISSIONED: 1941
DISPLACEMENT: 42,000 tons
SPEED: 30 knots
BELT ARMOUR: 32cm (12.6in) max.
ARMAMENT: 8 x 38cm (15in) +
    12 x 15cm (5.91in) guns

had old-fashioned secondary armament featuring separate surface and AA guns.

Japan's most modern battleships were bigger yet. The 65,000-ton *Yamato* and *Musashi* (two sister ships were planned, but not built as battleships) carried 9 x 460mm (18.1in) main guns, a mass of secondary weapons and had 406mm (16in) armour. Ironically, despite their status as the most powerful battleships ever built, these two vessels were among the few WWII battleships actually to succumb to the new menace of air attack while at sea.

The USA at War

The war was kinder to the United States than to any other warring nation. The country ended the war with the lowest casualty rate of any major combatant and the American people generally prospered in the economic boom the war brought.

In late 1940 President Roosevelt spoke of his wish that the United States would become the "arsenal of democracy". Even before Pearl Harbor this wish was fast becoming a reality and its consequences were transforming the USA and its relationship with the rest of the world. No informed observer in 1939 was in any doubt about the country's potential strength, but when the war in Europe began 15 per cent of the workforce was unemployed, factories were idle and other economic indicators confirmed the gloomy picture.

By 1945 much had changed. In the course of the war the USA manufactured a stunning

Above: Making aircraft engines was one of many new employment areas opened to American women.

total of 300,000 military aircraft, 86,000 tanks and vast amounts of every other conceivable kind of military equipment. The US armed forces were the most lavishly equipped in the world and, over and above their supplies, American production also met an estimated 25 per cent of

British needs, 10 per cent of Soviet requirements and large proportions for every other Allied power.

Some 15 million men served in the US armed forces during the war along with 350,000 women, a total only surpassed by the Soviets. From 1942 this huge military establishment was controlled from the Pentagon, the world's largest building, which had been newly opened in Washington, DC.

At the heart of the Pentagon was another new body, the Joint Chiefs of Staff, the committee of the armed service chiefs, which co-ordinated American planning and had an effective liaison organization working with its British equivalent. Although there were many disputes, Anglo-American planning was much more effective than the chaotic German or Japanese equivalents, with their capricious leaders and vicious inter-service rivalries. The JCS organization was truly a major factor in the Allied victory.

In December 1941 Congress passed the War Powers Act giving the President more executive authority than he had ever had before. A whole range of government agencies was soon set up to manage various important aspects of the war economy – the War Production Board and Office of War Mobilization being among the most important. Manpower was perhaps the first issue. Conscription

Below: The Chrysler plant in Detroit, producing M3 tanks in 1942.

Above: US rationing in 1941 was so lenient that this driver could fill a spare can as well as his tank.

JAPANESE-AMERICANS

In 1941 there were about 120,000 Japanese (over 60 per cent of them US citizens) living in the continental USA, and more than a third of the population of Hawaii was of Japanese descent. Though carefully watched, those in Hawaii were allowed to live normally. However, in early 1942, almost all Japanese living along the American West Coast were deported to a number of unpleasant detention camps inland and held there until late 1944. Fears of spying and sabotage were the official reason given, but the true motivation was a racist one. Canada also operated a similar policy for its Japanese population.

Above: Some 110,000 Japanese-Americans were sent to detention camps in 1942. German- and Italian-Americans were not similarly treated.

never dug as deep in the USA as in many other countries – married men were seldom drafted, for example. Men not drafted did not have to take war jobs and women were not compelled to work or serve in any way.

Big business and ordinary people both prospered during the war. Corporate profits soared and so did farm prices. Wages rose 50 per cent in real terms. The number of women working outside the home also expanded by about a third to 22 per cent of the workforce. Previously, women in paid work had generally been young, unmarried or childless, but older women and mothers commonly took jobs during the war.

Although the idea of women doing "man's work" was much publicized, the reality was slightly different. Few women moving into the workforce took over jobs previously done by men; rather they took new jobs, often of types that had not commonly existed previously.

The archetypal character Rosie the Riveter working in a shipyard scarcely existed, if only because shipbuilding was rapidly moving to the more modern welding method of construction. Indeed at the heart of the war production boom were a mass of such productivity improvements: new technologies, better machine tools, greater use of assembly-line methods and more.

Although many women were working and prospering, they were still paid about a third less than men in comparable jobs. A similar situation applied to African-Americans and other racially disadvantaged groups. Their wages rose faster than those of whites during the war, but there was still a large gap and racism remained ubiquitous both in civilian life and in the military. Military units were racially segregated and few non-whites were permitted to serve in combat formations, and fewer still became officers.

Japan's First Advances

Japan's initial naval successes were soon matched by triumphs on land. Malaya and Britain's great naval base of Singapore were conquered with ease by mid-February 1942 and the final Allied bastion in the Philippines lasted little longer.

Simultaneously with the attack on Pearl Harbor, Japanese forces began operations against the Philippines and Malaya. The Japanese landings in Malaya began at 01:00 hours on 8 December 1941 (local time). This was actually shortly before the 7 December raid on Pearl Harbor because of the effects of the International Date Line. The first air attacks on Luzon in the Philippines came a few hours later.

As well as these major targets, the Japanese also began a rapid campaign to conquer Hong Kong, while other forces quickly took control of Guam, Wake and other small US-held islands, despite an energetic defence by Allied forces.

MALAYA UNDER ATTACK

The 65,000-strong Twenty-fifth Army, led by General Yamashita Tomoyuki, began its campaign

Above: Although outdated by European standards, Japanese tanks like this Type 95, in action on Luzon, led their advance in 1942.

with landings on the South China Sea coast of northern Malaya and southern Siam (now Thailand). They were opposed by almost 90,000 British, Indian and Australian troops under the overall command of Air Marshal Robert Brooke-Popham, with General Arthur Percival as land commander. The Japanese quickly gained command of the

air and, after sinking HMS *Prince of Wales* and *Repulse*, had complete control at sea.

Many of the Japanese troops were battle-hardened veterans of the China war but, contrary to popular belief, they had had no special jungle training. They had a few tanks to break up any defensive positions they met and bicycles for mobility. By contrast the Allied troops were mostly badly trained and poorly led at almost every level. The result was that the Allied forces were hustled into retreat, time after time.

FALL OF SINGAPORE

By the end of January 1942 the Allied forces had withdrawn to Singapore island, over 950km (590 miles) from the initial Japanese landings. Singapore was supposedly an impregnable fortress, but its defences had been built with a naval attack in

SINKING OF FORCE Z

As relations with Japan deteriorated in late 1941, the British government decided to send naval reinforcements ("Force Z") to Singapore to deter Japanese action. Two battleships, the *Prince of Wales* and *Repulse*, were sent but a planned aircraft carrier did not go. The admiral in charge was one of the least air-minded in the Royal Navy; he made a blundering attempt, without air cover, to intercept Japanese landing forces off Malaya. Instead, on 10 December, the British ships were tracked down and sunk with little difficulty by Japanese land-based aircraft.

Above: *Prince of Wales*, *Repulse* and an escorting destroyer (nearest) seen from a Japanese aircraft.

mind and were not well suited to opposing an advance across the Johor Straits to the north.

On the night of 8–9 February the Japanese surged over and soon pushed the defenders back to the edges of Singapore city itself. General Percival decided to capitulate, though his troops (recently reinforced) greatly outnumbered their attackers.

In the whole Malayan campaign the Japanese lost fewer than 10,000 casualties. The Allies had a similar number of killed and wounded, but in addition some 130,000 went into a brutal Japanese captivity that many would not survive.

THE PHILIPPINES

Japan's conquest of the Philippines was almost as rapid. General Douglas MacArthur led the defending forces of 110,000 Filipino troops and 30,000 Americans, though inevitably many were dispersed around the Philippine archipelago. Equipment and training were poor but they did have over 200 supporting American aircraft. However, MacArthur's incompetence allowed many of these to be surprised on the ground hours after news of Pearl Harbor arrived. From then on Japan had control in the air. With no prospect of relieving US Navy forces arriving, as pre-Pearl Harbor plans had anticipated, Japan was also dominant at sea.

Troops from General Homma Masaharu's Fourteenth Army landed on the main island of Luzon on 10 December with larger forces following from the 22nd. MacArthur ordered the outmatched Allied troops to withdraw to the Bataan Peninsula area, which they did by early January 1942.

They held out there until 9 April, in part because some of the attacking Japanese troops had been withdrawn to take part in other phases of Japan's offensive. The fortress island of Corregidor offshore did not

Above: American and Filipino soldiers under fire in the defences of the Bataan Peninsula.

surrender until 6 May. Again many Allied troops were sent into a cruel captivity.

Meanwhile, MacArthur had been ordered to leave in mid-March to take command of Allied forces in Australia. On his departure MacArthur had promised: "I shall return."

Right: Surrender of Singapore, 15 February 1942. General Yamashita insists that General Percival signs without further delay.

Japan's Continuing Successes

Even before Malaya and the Philippines had fallen, Japanese forces were surging into Burma and landing throughout the Dutch East Indies. Their campaigns there would be just as rapid and successful as those that had gone before.

Japanese troops moved into Burma on 14 December 1941. An attack began in earnest on 20 January 1942 when General Iida Shojiro's Fifteenth Army launched two divisions toward Moulmein from across the border with Thailand. Allied forces fought effectively at first, retreating gradually. On 23 February a vital bridge over the River Sittang was blown up with many Indian troops on the wrong side. After that the retreat turned into a rout.

The Japanese were reinforced with troops from Malaya, now that campaign was over, and Rangoon fell on 8 March. Further Japanese troops then moved in from Thailand, and pushed the remaining British forces northward toward India. Chinese troops, commanded by an American, General Joseph Stilwell, joined in the battle but

Above: Apprehensive Burmese wait by the roadside as Japanese invaders pass by.

Below: Japanese Type 89 tanks on the advance in 1942. Lightly armoured and with a low-powered 57mm (2.24in) gun, the Type 89 was mainly used in China later in the war.

were also defeated. The last remnants of the Allied force reached India in early May.

VITAL OILFIELDS

The Dutch East Indies was the greatest prize in Japan's planned advance because of its substantial oilfields, mainly on Sumatra, and the significant production of metal ores and other important commodities. Japan's air and naval superiority was the key to their advance.

There were three main lines of attack, with various units from General Imamura Hitoshi's Sixteenth Army providing the principal ground forces. Sixteenth Army had first moved from Japanese possessions in the Caroline Islands to the southern Philippine island of Mindanao. In early January it struck toward Borneo and the Celebes, using both paratroops and forces landed by sea, before

Left: Japanese infantry capturing
the Yenangyaung oilfields in Burma
in April 1942. Gaining access to
foreign oil was a key Japanese aim.

THE DOOLITTLE RAID

In early 1942 US aircraft carriers began striking back at Japan with minor raids on various Pacific islands. In April a more ambitious plan, an attack on Tokyo, was carried out. On 18 April the carrier *Hornet* launched 16 Army Air Force B-25 bombers, commanded by Colonel James Doolittle, from some 1,000km (625 miles) offshore. Minor damage was caused in the Japanese capital and three other cities. The raid was a major boost to domestic morale in the USA and helped persuade Japan's leaders to carry out what proved to be their disastrous attack on Midway.

Below: A B-25 sets out for Tokyo. Taking off in a B-25 from a carrier was a remarkable feat, but no naval aircraft had the range needed for this mission.

leap-frogging south to Timor and eastern Java. At the same time other forces landed along Borneo's northern coast and by late February/early March were in Sumatra and western Java. Local Dutch and other Allied forces surrendered on 8 March.

NEW NAVAL VICTORIES

There were a number of naval engagements during the campaign, fought on the Allied side by Dutch, British, Australian and US Navy ships, led by the Dutch Admiral Karel Doorman. The Allied force was defeated on 27 February 1942 in the Battle of the Java Sea, and wiped out in follow-up engagements over the next two days.

Australian-ruled territory had also fallen. In January the Japanese captured Rabaul on New Britain (part of Australia's New Guinea mandated territory) and would build up a major base there. On 19 February they further demonstrated their naval superiority in the theatre when four of the aircraft carriers from the Pearl Harbor attack force led a devastating

raid on Darwin in Northern Australia. On 8 March Japanese troops landed at Lae and Salamaua in New Guinea proper, as the first stage in a planned advance to Port Moresby on Papua's south coast.

Having supported various of the Japanese landings, the carrier force then moved further west in March to attack Ceylon and British Indian Ocean trade. The main British and Japanese naval forces did not come into action against each other, but two British cruisers and a small aircraft carrier were sunk. The ports of Colombo and Trincomalee were heavily raided, while a subsidiary Japanese force sank numerous merchant ships in the Bay of Bengal.

Japan had achieved a bewildering succession of victories across a vast area, but ambitious generals and admirals planned still more advances. First they planned to attack New Guinea and then (stung by the American pinprick Doolittle Raid on the Japanese Home Islands) they would advance across the central Pacific to Midway.

Coral Sea and Midway

These two battles were the first naval engagements in history in which the opposing ships never came into visual contact. They confirmed that aircraft carriers were now the dominant weapons in naval warfare.

As May 1942 began, Japanese commanders planned to extend their conquests by landing at Port Moresby in southern Papua, from where their aircraft might reach targets in Australia. Warned by code-breaking information, Allied naval forces were sent to stop them.

Leading the Allied force were the US carriers *Lexington* and *Yorktown*, supported by American and Australian cruisers and destroyers. Opposing them were three main Japanese forces: the Port Moresby invasion force of troop transports and escorts; a covering force

KEY FACTS–MIDWAY

PLACE: Central Pacific, near Midway Island

DATE: 4 June 1942

OUTCOME: 4 Japanese aircraft carriers sunk; Japanese naval superiority lost.

THE BATTLE OF MIDWAY

As in other battles in the Pacific, the Japanese plan for Midway was over-complicated and failed to concentrate on the main objective.

including the small carrier *Shoho* and four cruisers; and a carrier strike force based around the *Zuikaku* and *Shokaku*. Also a small detachment was to land on Tulagi in the Solomon Islands to set up a base there.

CORAL SEA BATTLE

After preliminary skirmishes, the main action began on 7 May. US aircraft sank the *Shoho*; in return the Japanese carriers sank an American tanker and a destroyer. A full-scale carrier battle followed on the 8th. The Japanese came off better, sinking the *Lexington* and damaging

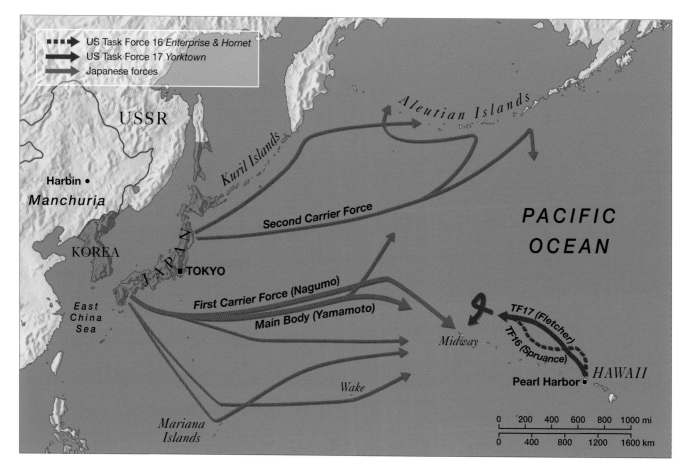

the *Yorktown* badly. However, the *Shokaku* was also heavily hit and many of the *Zuikaku*'s aircraft were destroyed. Neither ship would be able to join the Midway operation, which was scheduled next, and the Port Moresby invasion was called off.

THE MIDWAY PLAN

Admiral Yamamoto, principal Japanese Navy commander, had realized for some time that his victory at Pearl Harbor was incomplete unless he could destroy the US Navy's carrier force. He planned to capture Midway, far across the Central Pacific toward Pearl Harbor, confident that this move would force the Americans into battle.

The Japanese plan was characteristically complicated, with a selection of transport and covering forces closing in on Midway from various directions. Two smaller carriers and other major warships were to carry out a diversion against the Aleutian Islands, while the main forces included four carriers and a battleship and cruiser group.

Unfortunately for the Japanese, their intentions had once again been discovered by the US code-breaking services. The Americans were, therefore, able to ignore the Aleutians force (whose valuable carriers were thus wasted) and concentrate their strongest units near Midway before the Japanese knew they were in the area. The Americans deployed the carriers *Hornet* and *Enterprise* and supporting vessels, which were joined at the last minute by the rapidly repaired *Yorktown* (which the Japanese were convinced would be out of action for much longer).

Above: The Japanese carrier *Shokaku* under attack and on fire during the Coral Sea battle.

Above: A Japanese torpedo strikes home on the *Yorktown* at Midway, despite defensive anti-aircraft fire.

The main action began early on 4 June with Japanese air attacks on Midway Island. Thinking these attacks had not been successful enough and unaware for the moment of the American carriers' presence, Admiral Nagumo Chuichi, commanding the Japanese carriers, then ordered preparations for another attack on Midway.

While this was happening, attack aircraft from Midway Island and then from the American carriers arrived. Most were easily dealt with by the Japanese fighter screen, but a final dive-bomber group from the *Enterprise* inflicted crippling damage on *Akagi*, *Kaga* and *Soryu* (all three would sink later). The fourth Japanese carrier, *Hiryu*, sent a new strike that badly damaged *Yorktown*, but later *Hiryu* was itself fatally hit. Yamamoto abandoned the Midway operation.

Even though *Yorktown* was sunk on 7 June by a Japanese submarine, Midway had been a stunning American success. The backbone of the Japanese fleet was destroyed; not only the lost ships but also the many well-trained aircrew would not be readily replaced.

Below: A Devastator torpedo-bomber squadron ready for take-off on the *Enterprise* at Midway.

Aircraft Carriers

Airpower enthusiasts had been claiming since the 1920s that the aircraft carrier would soon supplant the battleship as the ultimate naval weapon. In the vast spaces of the Pacific Ocean, at least, this certainly proved to be true.

Experiments with aircraft-carrying ships were begun before WWI and the first warship to succumb to air attack was a German vessel sunk by a Japanese aircraft in 1914. But the first two true aircraft carriers with flight decks running from end to end did not enter service with Britain's Royal Navy until shortly after the end of the war.

By the time WWII began, Britain, the USA and Japan had significant aircraft-carrier forces, and built more as the war proceeded. The only other carrier in service in 1939 was France's experimental vessel *Béarn*, which did not ever see combat. Germany was building one and Italy later started two, but none of these was completed.

The three leading navies all developed ideas about carriers in the inter-war period. Both the

Below: The Independence-class light carrier USS *Langley*, with an Essex-class ship behind, at Ulithi atoll in 1944.

IJNS *KAGA*

The largest carriers in service at the start of WWII were, like *Kaga*, conversions of hulls originally planned as battleships or battle-cruisers just after WWI. A reconstruction in 1936 included the installation of an unusual downward-pointing funnel. *Kaga* participated in the Pearl Harbor attack and the raid on Darwin, but it was sunk at Midway.

COMMISSIONED: 1929
DISPLACEMENT: 38,000 tons
SPEED: 28 knots
AIRCRAFT: up to 90
GUNS: 10 x 200mm (7.87in) + 16 x 127mm (5in)

Japanese and the Americans headed in what proved to be the best direction. They realized that the carrier's best defence – and its best means of attack – were its aircraft and, therefore, the more the better.

Japan and the USA built carriers in which the hangar area under the flight deck was relatively lightly enclosed, with sides that could be opened for ventilation. This allowed the most room for aircraft.

Several of their early ships – Japan's *Akagi* and *Kaga* and the USA's *Lexington* and *Saratoga* – were particularly large, all well over 30,000 tons, and could carry up to 120 aircraft each. And with big air groups they also developed effective techniques for handling them during operations.

BRITISH DESIGNS

Britain had led the way in early carrier development but fell far behind in the inter-war years. The main reason for this was that the Royal Air Force – not the Royal Navy – controlled the supply of aircraft and pilots for maritime duties and these were given low priority. The Navy also chose to build carriers with "closed" hangar decks, which reduced aircraft capacity, although it did lead to the development of improved fire-control precautions.

In the late 1930s, aware that their obsolescent aircraft would be unable to protect the carriers

completely, especially in the Mediterranean where land-based aircraft would always be within striking range, the British took this process a step further by armouring the flight decks of their new carriers. This reduced capacity even more; HMS *Illustrious*, begun in 1937, initially carried only 36 aircraft, though capacity for *Illustrious* and its 5 similar successors was increased during the war.

PACIFIC CARRIERS

Both Japan and the US Navy went to war in 1941 with a number of rather smaller carriers in addition to the largest ships already mentioned. Most were around 18,000 tons and carried 70–80 aircraft. Ships in this category included *Yorktown*, *Enterprise*, *Hiryu* and *Soryu*. Other notable vessels included the 25,000-ton *Shokaku* and *Zuikaku*, and the somewhat smaller *Wasp* and *Ranger*.

Japan introduced a number of smaller carriers in 1941–2, several being conversions of

HMS *ARK ROYAL*

Probably the most famous British ship of the early war period, HMS *Ark Royal* was wrongly reported as sunk several times by German propaganda. *Ark Royal* mainly served in the Mediterranean (including in the action off Cape Spartivento, seen below) but a Swordfish from *Ark*

Royal also made the torpedo hit that crippled *Bismarck*. *Ark Royal* was finally sunk by a U-boat in November 1941.

COMMISSIONED: 1939
DISPLACEMENT: 22,000 tons
SPEED: 30 knots
AIRCRAFT: up to 60
GUNS: 16 x 4.5in (114.3mm)

other types of vessel. In all Japan completed 17 carriers (both large and small) during the war. None of these played a

substantial part. This was not because they were all inadequate ships but rather that, by the time they came into service, Japan's cadre of trained naval aircrew had been wiped out and could not be quickly replaced.

The US Navy had no such problems. Its 27,000-ton Essex-class ships could carry over a hundred aircraft; 24 saw service from early 1943 onward. In addition the US Navy had the 9 ships of the Independence class of light carriers, some 10,000 tons and carrying around 40 aircraft. With ample numbers of well-trained pilots and excellent aircraft, it was these ships that led the US advance across the Pacific to defeat Japan.

Left: USS *Essex*, lead ship of its class, with about 50 of its large air group parked on deck.

The Turn of the Tide

*Desperate struggles in the Solomon Islands and on New Guinea from mid-1942
saw the Japanese advance turned back for the first time. The Allies gained the
upper hand in vicious jungle fighting and a series of dramatic naval battles.*

Having overrun the East Indies so easily, Japan then planned to capture bases in southern Papua, as a prelude to a possible attack on Australia and others in the Solomon Islands, to threaten communications between Australia and the USA. As well as Australian and American ground and air forces, the resulting battles would draw in the main strength of both the Japanese and US Navies.

NEW GUINEA

After Japan's naval attack was thwarted in the Battle of the Coral Sea, General Horii Tomitaro's South Seas Detachment began an advance in July 1942 from the north coast of Papua over the precipitous Owen Stanley Range on the Kokoda Trail toward Port Moresby. Despite the terrible conditions (for both sides) the Japanese soon pushed back the weak Australian forces, now part of General MacArthur's Allied South-West Pacific Command. Allied air power and some reinforcements halted Horii on the approaches to Port Moresby in September and the Japanese retreated (so short of supplies some resorted to cannibalism).

In the meantime Allied air and naval forces had begun advances round the eastern tip of Papua, beginning at Milne Bay in August. By November Australian troops attacking along the Kokoda Trail and American troops moving along

Above: A wounded Australian near Buna is helped by a Papuan, one of many who assisted the Allied forces.

COAST WATCHERS

The Australian Navy used Coast Watchers, people who provided vital information to Allied forces during the Solomons campaign and other battles. The service had been set up before the war to cover a range of locations across the little-inhabited South-west Pacific. From their lonely and isolated posts, Coast Watchers reported movements by air and sea and also supplied weather reports, which was often just as important. Though they had to defend themselves against Japanese patrols, their role was not to fight. Admiral Halsey said that Guadalcanal would not have been won without their help.

the north coast had closed in on the well-fortified Japanese beachheads at Buna and Gona. The remaining Japanese forces there were wiped out by the end of January 1943.

Guadalcanal Island, the focus of the other flank of the campaign, saw conditions that were just as horrible. Japan had set up a seaplane base at nearby Tulagi in May and a small force was working on an airfield on Guadalcanal itself. The 1st US Marine Division landed on Guadalcanal on 7 August 1942, captured the airfield and put it into Allied service as Henderson Field. At first the Marines were not well supported from the sea; Japanese reinforcements began attacking the Marines' beachhead. During the six-month struggle that followed the Japanese troops fought fiercely, but their attacks were not well co-ordinated and were beaten off after a series of tough battles. Japan evacuated the island in early February 1943.

NAVAL BATTLES

Both sides' efforts to supply and reinforce their troops led to six major naval battles and a range of smaller engagements. At first the Japanese had the upper hand. In the early hours of 9 August, in the Battle of Savo Island, night-fighting skills and superior torpedo equipment helped a group of Japanese cruisers and destroyers sink four opponents with little loss.

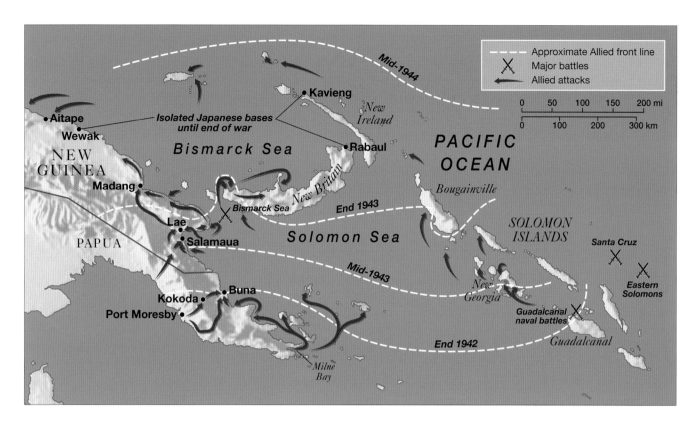

THE SOUTH-WEST PACIFIC

From summer 1942, Allied forces conducted a two-pronged advance along the north coast of New Guinea and up the Solomon Island chain.

On 23–4 August the Battle of the Eastern Solomons between the main carrier forces was a narrow Japanese victory; and in subsequent weeks two US carriers were sunk by submarines. However, by then aircraft from Henderson Field dominated the waters around Guadalcanal by day; but at night Japanese warships under Admiral Tanaka Raizo, known to the Marines as the "Tokyo Express", had the ascendancy.

A second carrier battle on 26 October, the Battle of Santa Cruz, was inconclusive, but the tide turned with the Battle of Guadalcanal taking place on the

nights of 12–13 and 14–15 November. By then the American forces were learning to use their superior radar equipment to win night battles. The fighting between the warships was by no means one-sided but several Japanese troop transports were sunk. The Japanese had the better of another night encounter (the Battle of Tassafaronga) in late November. However, by then the attrition of their naval forces had become so severe that in early January the Tokyo Express finally went into reverse – to evacuate rather than supply the island. The Allied counter-offensive in the Pacific was now well under way.

Right: A Japanese bomb explodes aboard the *Enterprise* during the Battle of the Eastern Solomons.

Light Cruisers

Capable of giving a good account of themselves in battle, light cruisers were maids of all work, but they also were used for long-range trade protection missions, shore bombardments and other varied duties.

Cruisers carrying guns of roughly 150mm (5.91in) were probably the most ubiquitous warships of the conflict. They formed a vital part of naval task forces, both large and small, in every campaign.

WWII light cruisers can be divided into three sub-types: larger ships usually displacing at least the Washington Treaty limit of 10,000 tons or more and carrying 12 or 15 main guns, usually in triple turrets; smaller general-purpose ships of 6,000–8,000 tons, armed usually with 8 main guns in twin turrets;

Below: USS *Atlanta* (San Diego class) served in the South Pacific in 1942 but was sunk off Guadalcanal.

and anti-aircraft cruisers of 6,000 tons, armed with 10–12 smaller dual-purpose guns.

LARGE LIGHT CRUISERS

This type developed largely as a result of the pre-war arms race in the Pacific, beginning with Japan's construction of the Mogami class in the 1930s. With an impressive top speed of 35 knots, and 15 x 155mm (6.1in) guns, they supposedly had a relatively modest displacement of 8,500 tons (but, in fact, they were well over 10,000 tons).

The US Navy replied with its Brooklyn class and Britain with the Southampton class. Their wartime successors were the Cleveland (US) and Edinburgh

and Fiji (UK) classes, respectively. They were all around 10,000 tons and built with 12 x 6in (152mm) guns in triple turrets (except the Brooklyns, which had 15). During the war some ships had a 6in turret removed and replaced with smaller-calibre AA weapons.

Most of Japan's large cruisers were in the heavy-cruiser category and by Pearl Harbor, the Mogamis had also been rearmed with 203mm (7.99in) guns.

SMALL LIGHT CRUISERS

Although the US and Japanese navies had a number of cruisers of this type dating from the 1920s or earlier, including ten ships of the US Omaha class,

most vessels in this category served with European navies. This was the type of cruiser that Britain's Royal Navy most wanted to build in the inter-war period, with a good combination of long range and effective gunpower. The Leander class of the mid-1930s, for example, carried 8 x 6in guns and could make 32 knots on a displacement of some 7,200 tons. France, Germany and Italy all had comparable ships.

However, the most modern German ship, in service from 1935, was the *Nürnberg*, which

carried 9 x 15cm (5.91in) guns and was otherwise similar to the ships of the British Royal Navy.

The French and Italian Navies had slightly different priorities. In the 1930s the Italians built very fast ships (and made them seem even faster by falsifying the results of their trials) and the French replied in kind. The Italian Duca d'Aosta class (8,500 tons, 8 x 152mm/ 5.98in guns), for example, supposedly reached 37 knots.

AA CRUISERS

Many of the cruisers described above had relatively limited AA capability because their main gun mountings and control systems were only suitable for use against surface targets. As the war progressed, ships in all navies gained additional light AA guns, but Britain and the USA also saw a need for cruisers with greatly enhanced heavy AA firepower. The US Navy's San Diego and Britain's Dido and Bellona classes all carried large batteries of dual-purpose main guns, 10 x 5.25in (133mm) for the Didos and 12 or even 16 x 5in (127mm) for the San Diegos. All were also fast ships, capable of 33 or 34 knots to keep up with fast aircraft carrier forces.

Above: The *Giuseppe Garibaldi* carried 10 x 152mm (5.98in) guns and could reach a reported 34 knots.

KÖLN

The German Navy built three K-class cruisers in the late 1920s. Unusually, six of the nine main guns were aft. In service *Köln* and its sisters were found to lack stability in heavy weather. In 1940 *Köln* participated in the invasion of Norway but saw little action thereafter before being sunk by air attack in port in 1945.

COMMISSIONED: 1930
DISPLACEMENT: 6,700 tons
SPEED: 32 knots
AIRCRAFT: 2 x Arado 196
ARMAMENT: 9 x 15cm (5.9in) + 6 x 8.8cm (3.46in) guns; 12 x 533mm (21in) torpedo tubes

HMS *AJAX*

One of the five British Leander-class light cruisers, *Ajax* had an eventful war career. In 1939 *Ajax* joined two other cruisers, *Exeter* and *Achilles*, in hunting down the *Graf Spee*. In 1944 *Ajax* was one of the ships in the D-Day bombardment force. This record in action proved the usefulness of the relatively small cruisers of this type.

COMMISSIONED: 1935
DISPLACEMENT: 7,200 tons
SPEED: 32 knots
ARMOUR: 102mm (4in) max.
ARMAMENT: 8 x 6in (152mm) + 8 x 4in (102mm) guns; 8 x 21in (533mm) torpedo tubes

Japan's Brutal Empire

Japan's military conquests were supported by an authoritarian military government at home and ruthless exploitation of the people and resources of the conquered territories, though this was disguised by anti-imperialist propaganda.

In 1938 Japan's Prime Minister, Prince Konoye Fumimaro, spoke of Japan's aim to create a "New Order" in Asia. In 1940, after a period out of office, Konoye returned as premier and announced Japan's plan to establish a Greater East Asia Co-Prosperity Sphere. These ideas, and their slogan "Asia for the Asiatics" formed part of the backdrop to the creation of Japan's wartime empire. European and American racism and imperialism were to be rejected in favour of a vague commonwealth from which all Asians were to benefit – ideas that were designed to appeal both at home in Japan and throughout Asia.

MILITARISM AT HOME

From the later 1920s Japan itself became an ever more militaristic and rigorously regimented

Above: A Japanese soldier proudly displays the severed head of a Chinese man, Shanghai, 1937.

Below: From 1941 Japanese high-school students had compulsory military training in the curriculum.

society. The armed services took effective control of the government. As well as expanding Japan's war in China, factions within the army had no hesitation in assassinating opponents or planning military coups. The economy was increasingly industrialized and militarized, with businesses and trades unions also brought under government control.

The education system emphasized military and nationalistic values: for example, maps in school textbooks showed much of South-east Asia as rightfully forming part of Japan's Empire. And in the background were the Tokko, or Special Higher Police (usually referred to in English as the "Thought Police"), who used torture and other repressive methods to ensure citizens behaved as the government wished.

CRUELTY ABROAD

The reality of Japan's rule overseas belied the intent expressed in Konoye's slogans. Most obvious to the Western Allies was its treatment of prisoners of war. Japanese troops were taught that it was disgraceful to surrender and regarded with contempt any enemy who did so. Many American prisoners, captured in the Philippines in 1942, died of ill-treatment on the so-called Bataan Death March and around 12,000 British and Australians were starved, beaten and worked to death on the

Above: Ho Chi Minh, leader of the anti-Japanese resistance movement in Indo-China, was helped by US special forces later in the war.

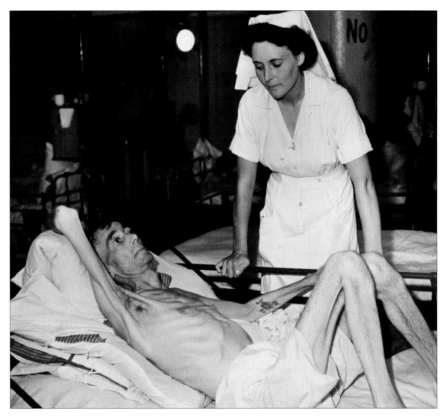

Above: An emaciated British prisoner receives medical treatment after his liberation from Japanese-occupied Hong Kong in 1945.

Siam–Burma railway line, to give only two of numerous notorious examples.

However, if only because relatively few Westerners came into Japanese hands, these brutalities bear little comparison with those inflicted on other Asians – for example, as many as 90,000 Malaysian, Thai and other Asian labourers died on the Burma Railway. Japan's attitude to other Asians was more brutal, racist and exploitative than any of the much-despised colonial powers.

The outside world first became aware of this with the infamous "Rape of Nanking", in December 1937, which was widely reported elsewhere. When Japanese troops captured the town (now usually called Nanjing), they went on an extended rampage of murder, rape, looting and arson, which killed a quarter of a million people (according to some estimates). Throughout the war in China, Japanese tactics made no distinction between civilians and military opponents; the watchwords were the "three alls" – kill all, burn all, loot all.

Other Japanese methods were more insidious. In the puppet state of Manchukuo, for example, the Japanese authorities encouraged and greatly expanded the production of opium. They also pushed users of opium there and in the rest of China to switch to the more dangerous morphine and heroin compounds, as part of a deliberate strategy to keep the indigenous population docile. The revenues from the scheme were used for the benefit of Japan's Kwantung Army.

For all that Japanese rule was far harsher than the colonial regimes that had preceded it, it also discredited such pre-war governments. In the aftermath Asia would soon come to be ruled by Asiatics to an extent unimaginable before.

COMFORT WOMEN

One of the nastiest aspects of Japanese rule was the imposition of forced prostitution on many thousands of mainly non-Japanese women. In part to prevent atrocities like the mass rapes that had occurred at Nanjing, the Japanese Army set up official military brothels in China and elsewhere. Up to 200,000 women worked in these establishments and many of them were teenagers. They were supposed to receive payments, but many were forced into unpaid prostitution. Most of the women were Korean.

The Holocaust Emaciated survivors shortly after the liberation of Buchenwald camp in 1945.

Infantry Weapons An American infantryman training with his M1 Garand semi-automatic rifle.

US Strategic Bombing, 1942–3 A B-24 Liberator falls in flames after being hit by flak.

ICELAND

ATLANTIC OCEAN

	Incorporated in Hitler's Reich
	German-occupied 23 June 1944
	German allies 1942–44
	Neutrals
——	Front line 5 Dec 1941
- - -	Front line May 1942
-·-·-	Front line 18 Nov 1942
-··-··-	Front line 4 July 1943
······	Front line 23 June 1944

0 100 200 300 400 500 mi
0 200 400 600 800 km

NORWAY

SWEDEN

FINLAND

North Sea

Baltic Sea

U S S R

UNITED KINGDOM

IRELAND

ESTONIA

LATVIA

LITHUANIA

DENMARK

EAST PRUSSIA

NETHER-LANDS

POLAND

BELGIUM

GERMANY

CZECHOSLOVAKIA

FRANCE

SWITZER-LAND

AUSTRIA

HUNGARY

ROMANIA

YUGOSLAVIA

Caspian Sea

PORTUGAL

SPAIN

Corsica

ITALY

ALBANIA

BULGARIA

Black Sea

8 Oct 1943

Sardinia

Mediterranean Sea

Sicily

17 Aug 1943

GREECE

TURKEY

IRAN

15 Jan 1943

SYRIA (French)

CYPRUS (British)

IRAQ (British)

15 Jan 1943

TUNISIA (French)

PALESTINE (British)

ALGERIA (French)

18 Nov 1941

31 Aug 1942

TRANS-JORDAN (British)

15 Jan 1943

1 Jan 1942

North Africa front lines

LIBYA (Italian)

EGYPT (British)

THE GREAT STRUGGLE IN EUROPE

Just as his forces were being pushed back from Moscow in December 1941, Hitler assured his eventual defeat by declaring war on the USA. By the middle of 1944 Axis troops had been thrown out of the USSR and expelled from Africa; Italy had made peace; Anglo-American bombers were pounding Germany from the air; and Allied land forces were poised to return across the Channel in the Normandy invasion. Yet, between these landmarks, there lay many hard-fought campaigns and much dreadful suffering.

This was the period when the Nazi murder campaign against Europe's Jews was at its height. The Eastern Front saw the horrific sieges of Leningrad and Stalingrad and the war's greatest tank battle at Kursk. The Western Allies gradually grew in confidence and strength, initially in their victories in North Africa, then in their burgeoning power at sea and in the air that made first Hitler's Kriegsmarine, and then the Luftwaffe, shadows of the forces they had once been.

In his post-war memoirs Churchill wrote that the Allied victory was inevitable from the moment the USA joined the war. In hindsight this may indeed have been true, but it took the vicious battles of 1942–4 to turn this judgement into a reality.

Assault Guns An American soldier examines a German Jagdpanther knocked out in 1945.

Clearing the Ukraine Red Army troops being welcomed warmly on their recapture of Kiev.

Naval Weapons and Electronics Guns on the foredeck of the battleship USS *Iowa*.

To Stalingrad and the Caucasus

Despite their losses in the first terrible winter of war on the Eastern Front, the Germans were ready to attack again in 1942. Once more they used their superior tactical skills to win early successes, but a final decisive victory remained elusive.

After pushing the Germans back from Moscow with their December 1941 counter-offensive, the Soviets tried to develop a general advance all along the Eastern Front. Although these attacks created large bulges into German-held territory, the major German defensive positions held with relatively little difficulty. By the time the spring thaw brought operations to a halt, it was clear that the Germans would attack again in strength that summer.

Although the German forces had been substantially reinforced by troops from Hungary, Romania, Italy and even Spain, they no longer had the power to attack over the whole Eastern Front. Instead Hitler decided to concentrate in the south. He chose two objectives: a drive into the Caucasus as far as the

Above: Romanian troops in action in the Crimea alongside their German allies in June–July 1942.

Below: German infantry advancing during their successful counter-offensive near Kharkov in late May.

Caspian coast to gain control of the oilfields there, and an advance to Stalingrad (Volgograd) to establish a protective front along the Don River north to Voronezh.

PRELIMINARY ATTACKS

Late spring and early summer saw new German successes. Attacks in the Crimea led by General Erich von Manstein wiped out Soviet forces on the Kerch Peninsula in May and took Sevastopol by early July. To the north a Soviet attack near Kharkov in mid-May was crushed. Together these operations saw a further 450,000 prisoners fall into German hands.

The main German attacks began on 28 June. Voronezh was captured within days as the Soviet front at first fell apart before the German attacks. By late July other German attacks captured Rostov and opened the way south into the Caucasus.

However, Hitler was already changing his mind on priorities and readjusting his forces. In mid-July General Friedrich Paulus's Sixth Army had been told to rush east toward Stalingrad and much of the tank forces of Fourth Panzer Army, which had been meant to lead the drive into the Caucasus, were switched to join the right flank of Sixth Army's advance. To complete the range of contradictory objectives, Hitler also ordered Manstein's victorious troops from the Crimea to

THE GERMAN ADVANCE TO STALINGRAD

German advances in the summer of 1942 greatly lengthened their front line but made no decisive gains.

Above: German troops at Maykop in late summer 1942. Retreating Soviets have set the oil wells alight.

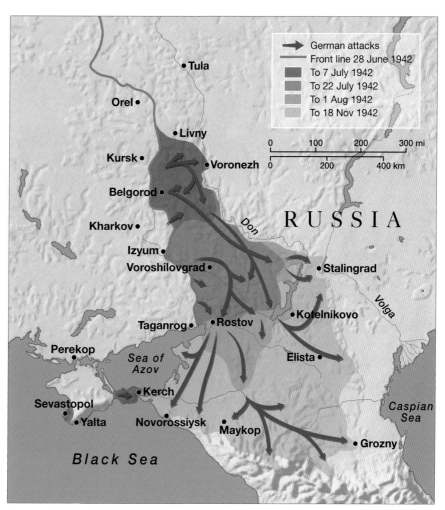

move north to Leningrad to finish the destruction of the Soviet resistance there.

SOVIET RESPONSES

For his part Stalin had misjudged the direction of the German attacks, fearing an advance toward Moscow from the south and, accordingly, had not deployed his reserves well.

However, the general Soviet organization had greatly improved. New production was being used to create air and tank armies that might provide a proper answer to German power in these respects, while more authority was being given to commanders to control their operations without political interference from their unit commissars. Finally, a group of ruthless and efficient generals was becoming established in senior positions, who would make good use of the ever-stronger Soviet forces.

For the moment the German advance continued. By early August, Sixth Army was destroying the Soviet forces in the Don bend, west of Stalingrad. By the end of the month it had reached the Volga, linked up with Fourth Panzer Army and closed to within a handful of kilometres of the city. Meanwhile, with supplies and air support having been switched to the Stalingrad sector, the advance in the Caucasus had ground to a halt, short of most of its objectives and never to be resumed. For Hitler and for Stalin, capturing or holding Stalingrad was now the only objective that mattered.

Below: A German tank unit halted in the vast and featureless steppe, south of Stalingrad, is overflown by a liaison aircraft.

The Holocaust

In a war of many atrocities one series of actions stands out as particularly horrific – Nazi Germany's attempt, at Hitler's command, to murder all the Jews of Europe solely because of their race and religion.

There are aspects of the vast history of WWII in which a simple factual account seems inadequate. The list of Jews murdered by the Nazis – at least 2.5 million Poles, 750,000 from the USSR, almost as many each from Hungary and Romania, and tens of thousands more from almost every occupied country in Europe – cannot begin to explain the suffering and vile cruelties involved. At the same time, however, failing to state what happened and how it came to happen can only assist those who would downscale or even deny the nature of these crimes.

EARLY PERSECUTIONS
Although Jews in pre-war Germany, and in occupied western Europe in 1940–1, were viciously persecuted, there were few killings. Eastern Europe presented, from the Nazi point of view, a different

Above: Jewish families are taken from the Warsaw ghetto, probably to the Treblinka death camp.

Left: An *Einsatzgruppe* killer at work, murdering Jews at Vinnitsa in the Ukraine in 1941–2.

problem. They aimed to make their domains "Jew-free" and here there were many more Jews to deal with. During 1940 most Polish Jews were forced to live in "ghettos" in major towns. These were deliberately made overcrowded and insanitary and a typical food ration was less than 200 calories per day. By the middle of 1941 the half-million Jews in the Warsaw ghetto were dying at the rate of 2,000 daily.

Even this did not satisfy the Nazis. The armies invading the USSR were closely followed by *Einsatzgruppen* – SS murder squads whose mission was to kill Jews and communists. By the end of 1941 at least 600,000 Jews had been rounded up, shot and tumbled into mass graves. This was done quite openly, often watched and photographed by German Army soldiers and other personnel.

THE FINAL SOLUTION
By this stage leading Nazis were looking for a "final solution to the Jewish problem", by which they meant finding an easy way to murder all of the Jews in Europe. The process began in the autumn of 1941 and was put into high gear by a meeting of top Nazis – known as the Wannsee Conference – in January 1942.

The killing mostly took place in specially established death camps in occupied Poland.

From late 1941 Jews from the Polish ghettos were transported to the camps and gassed, followed later by transports from the rest of German-controlled Europe. Most of the camps killed Jews as soon as they arrived but the largest, Auschwitz-Birkenau, sent only a portion to the gas chambers immediately and worked most of the remainder to death in a range of factories. Survivors from the camps were marched west in 1944, as the Red Army approached, to endure yet more slave labour. A few emaciated victims were liberated in 1945.

Most of the Jews of Germany, Poland and the western USSR died, but elsewhere the picture was less uniform. The Bulgarian government refused to allow any Jews to be sent to the camps, Mussolini's regime largely left them alone, and almost all of Denmark's Jews were smuggled to neutral Sweden, for example.

In all countries many people risked their lives to help Jews, both on a small and larger scale. By contrast some in France's Vichy government willingly assisted in the deportation process, and most countries had collaborators who helped the Nazis in their dirty work. Many Germans later denied knowing anything of what had been done in their name (indeed the Nazis tried to keep secret what was happening in the camps), but the official Nazi explanation that the Jews were taken east for "resettlement" cannot really have fooled many people.

Right: Adolf Eichmann, the callous bureaucrat who managed much of the "Final Solution".

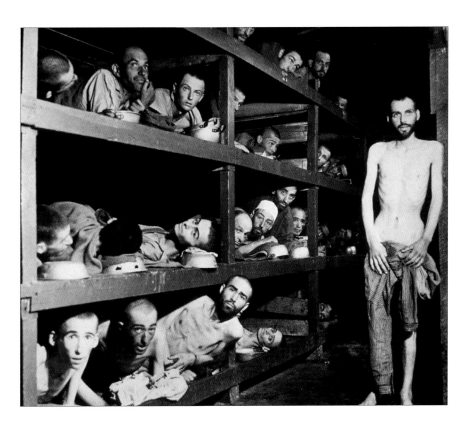

Britain and the USA had a fair idea of what was happening from the middle of 1942 but, other than a few protests, did nothing that might have slowed or halted the murders. Although since blamed on anti-semitism, this failure probably arose because it was so difficult to accept the appalling truth that 6 million Jews were being killed.

Above: Some of the survivors liberated from Buchenwald in 1945, including the Nobel Laureate Elie Wiesel, middle bunk, seventh from the left.

EUTHANASIA

Many of the techniques of mass murder used against the Jews had been developed for use in the Nazi euthanasia programmes. During the 1930s over 300,000 carriers of hereditary mental illnesses or disabilities were sterilized. From 1939 onwards, over 100,000 "inferior Germans" were callously murdered, many by the medical staff of the hospitals "caring" for them. Victims included senile patients, the insane and infants with physical handicaps or genetic conditions such as Down's syndrome.

Field Artillery

Despite the more spectacular contributions of tanks and aircraft, World War II was a war dominated by artillery – more than half of all casualties on all battle fronts came from the ever more deadly concentrations of artillery fire.

The backbone of every army's firepower came from the medium-calibre guns of the field artillery units. Divisions invariably included an artillery component dedicated to the support of the division's units, usually on the basis of roughly one artillery battery (of perhaps six guns) for each infantry or tank battalion in the formation. Guns in the field artillery class were usually of 75–105mm (2.95–4.13in) calibre and fired shells, weighing 10–15kg (20–35lb), to a range of 12–15km (7.5–9.5 miles).

TYPES OF WEAPONS

A typical weapon of this type was the US Army's standard M2A1 105mm (4.13in) howitzer, a model that had been in service since 1934 and remained in use well into the Vietnam era. In common with most similar weapons, the M2A1 could fire a variety of different types of shell, including high explosive (HE), high explosive anti-tank (HEAT), white phosphorus, smoke, and even a leaflet-carrying type. A variety of propellant charges was also supplied for range adjustments.

Germany's standard weapons, the 10.5cm leFH 18 and slightly modified leFH 18/40, were essentially similar. Britain's main field gun – the 25pdr, with a calibre of 3.45in (87mm) – delivered a slightly smaller shell. However, it lost nothing in range and compensated by being very

Above: A German 10.5cm (4.13in) leFH 18 gun on the advance in the Ardennes in late 1944.

quick firing. Soviet divisional artillery weapons were a mix of smaller guns still (76.2mm/3in M1936, 1939 or 1942 guns and others) and heavier 122mm (4.8in) types (M1931 guns and M1938 howitzers and others). The 122mm howitzer had a similar range to the field guns discussed above and fired a 21.8kg (48lb) shell.

The other major Axis powers, Italy and Japan, had various 75mm (2.95in) and 105mm weapons, which were comparable in performance to the types noted above. However, neither nation used its artillery very effectively in action.

ORGANIZATION

Surprisingly, armies generally used fewer artillery weapons in WWII than in WWI. This was mainly because artillery tactics had changed and the techniques used for controlling artillery fire had advanced greatly.

Much of the artillery fire of WWI had been devoted to preparatory bombardments – programmes of shelling in advance of a battle designed to smash enemy positions and disrupt enemy forces. Germany and the Western Allies laid much less emphasis on this type of action in WWII, recognizing that it was of limited effectiveness and could often be counter-productive. Instead they emphasized neutralizing fire while an action was actually

Right: British artillerymen in Burma in action with a 3.7in (94mm) pack howitzer. Dating from WWI but with an improved mounting, the gun could fire a 9kg (20lb) shell.

occurring. The aim was to suppress enemy firepower and ability to manoeuvre.

Although the Red Army deployed a great mass of artillery its use tended to be less sophisticated, especially in the first couple of years of war. Battery commanders might well be the only personnel able to make the calculations needed for more complicated fire plans – their juniors might even be illiterate and have no watches for timing any switches of target.

Communications also played a vital part. Western artillery batteries routinely used forward observation officers (FOO), equipped with radios to direct and adjust their fire, and they had elaborate inter-connections between artillery units. An FOO could call for fire not only from his own battery but also, on occasion, from as many as several hundred other guns within a very short time. This process was assisted by the development of various standard patterns and timetables of fire that could very quickly be put in place by numerous artillery units. The superior British and American artillery organization established by 1943–5 was a significant factor in the Allied victory.

M3 105MM HOWITZER

The M3 105mm howitzer was a lighter version of the standard M2 field gun fitted with a shorter barrel. It was intended for use by airborne forces and was also employed by support companies of infantry units (as shown here in New Guinea in 1943–4). It had roughly two-thirds the range of the parent gun but fired similar ammunition. Some 2,500 were made.

CALIBRE: 105mm (4.13in)
WEIGHT: 1,135kg (2,500lb)
LENGTH: 3.94m (12ft 11in)
RANGE: 7,600m (8,300yd)
SHELL WEIGHT: 15kg (33lb) HE

Stalingrad

The Battle of Stalingrad is rightly regarded as the turning point of WWII. Before then any defeats suffered by Germany were little more than minor setbacks; after the Stalingrad calamity, rant at his generals as he might, Hitler had lost the war.

As the German advance ground toward Stalingrad through the summer of 1942, the war came to the local people in earnest. Stalin at first forbade any evacuation of the civilian population to prevent any suggestion that the city was likely to fall. Work continued in factories that produced the weapons that could be used immediately. T-34s from the former tractor factory would be driven straight into battle, unpainted and often lacking gunsights. Men not already in the army were formed into militia units and sent, untrained, straight into action with predictable horrendous

Below: German infantry armed with submachine-guns prepare to attack amid the ruins of Stalingrad.

KEY FACTS

PLACE: Stalingrad

DATE: 12 September 1942– 2 February 1943

OUTCOME: German Sixth Army failed to capture the city and was itself annihilated.

---

casualties. Women and children toiled to dig trenches and build defences; other women crewed anti-aircraft guns.

ATTACKING THE CITY

On 12 September the first German troops moved into the city itself. The Germans chose to attack straight into the urban

area and soon became embroiled in a vicious house-to-house battle. Soviet tactics were to establish front-line positions as close as possible to the German lines – the safest place from German air and artillery attack.

At the end of August Stalin had sent his top general, Georgi Zhukov, to supervise the whole southern front and now a new commander, Vasili Chuikov, took over 62nd Army in the city itself. While Zhukov prepared for an eventual counter-offensive, Chuikov drove his troops relentlessly into combat. A steady stream of reinforcements was ferried under fire across the Volga into the ruins, just enough arriving (despite repeated crises) to keep at least part of the city in Soviet hands.

Above: General Vasili Chuikov, the tough commander of 62nd Army in the defence of Stalingrad.

"NOT ONE STEP BACKWARDS"

Underlying the desperate defence of Stalingrad was the brutal and merciless disciplinary system of the Red Army. In August 1941 Stalin had issued an order that anyone trying to surrender should be shot on the spot. On 28 July 1942 his Order 227 (known as "Not One Step Backwards") reinforced this message. Armies were to form a second line of troops behind an attack to shoot anyone who wavered. Officers who "allowed" their men to desert or troops cut off by the Germans but who later returned to Soviet lines were sent to punishment battalions, a virtual death sentence. At least 400,000 died in such units in the course of the war.

Above: Last survivors of the German Stalingrad garrison march off into Soviet captivity.

Soviet losses were huge, but German casualties mounted, too. At the end of September, and again in mid-October, the Germans came close to victory, but the defenders just clung on in a small part of the city's northern factory district.

GERMAN VULNERABILITY

The advance to Stalingrad had brought other consequences for the Axis forces. In the course of the summer Hitler had sacked several of his top generals after quarrels about the strategic plan. Now he himself took many command decisions, both big and small. The advance to Stalingrad had greatly lengthened the German front and, as a consequence, two Romanian armies and an Italian army had been brought into line on the flanks of Sixth Army and Fourth Panzer Army. The Romanian troops were poorly motivated and trained and very badly equipped. They would be the first targets for the planned Soviet counter-offensive.

The Soviets attacked north of Stalingrad on 19 November and to the south the next day. The Romanian fronts were shattered. Within a week the two advancing Red Army forces had joined up and cut Stalingrad off. General Paulus asked for permission to break out – and at that stage could almost certainly have done so – but Hitler refused, promising to supply the trapped troops by air. General Erich von Manstein was brought in to lead a relief attempt that nearly reached the surrounded pocket, but he had to retreat just before Christmas.

In the meantime the air supply effort to Stalingrad's frozen airfields had cost the Luftwaffe hundreds of transport aircraft and delivered only a fraction of the necessary supplies. Short of fuel and ammunition and slowly starving, Sixth Army now faced superior forces on all sides. The end was inevitable. The last German troops in the city surrendered on 2 February 1943.

Above: German Stukas in action over Stalingrad in the early stages of the siege, autumn 1942.

Infantry Weapons

At its most basic level, combat in land warfare depended on the qualities of the infantry soldier's personal weapons. US Marines (like soldiers everywhere) were taught: "My rifle is my best friend. … Without my rifle I am useless."

The most common personal weapons for soldiers in all armies of World War II were hand grenades, submachine-guns and rifles.

Hand grenades were essential in every close-quarter engagement in all theatres of war and were used in vast numbers. Some grenades had offensive and defensive versions, the former relying solely on blast effects, the latter producing splinters or shrapnel in addition. Special-purpose grenades also included smoke and incendiary

Left: A German combat engineer clears a path through some barbed wire during an exercise. He is armed with a Stielhandgranate 24.

types. In outward appearance there were two main kinds: egg-shaped varieties like the British No. 36, based on the WWI Mills bomb, or stick designs like the German Stielhandgranate 24.

RIFLES

In 1939, as for decades before, the standard weapon for soldiers in all armies was a magazine-fed single-shot rifle, firing a bullet of roughly 7.7mm (0.3in) calibre. Such a round was lethal, and in theory could be fired accurately, to ranges well over 1,000m (1100yd), but it was an unlucky casualty indeed who was struck by a deliberately aimed shot at even a third of that distance.

The best such weapon was the USA's standard rifle, the 0.3in (7.62mm) M1 Garand, which had the advantage of being a semi-automatic design. Most other major rifles were of the older bolt-action type. The British 0.303in (7.7mm) Lee Enfield No. 4 was perhaps the best of these because its mechanism could be operated most speedily. However, other types like Germany's 7.92mm (0.312in) Mauser 98K, were also sturdy, accurate and reliable.

Left: An American recruit is instructed in marksmanship using an M1 Garand 0.3in (7.62mm) rifle.

SUBMACHINE-GUNS

Only during the final months of trench combat in WWI were submachine-guns used. They had the advantage over rifles of a far higher rate of fire (ammunition supply permitting), but they were difficult for even a well-trained soldier to fire accurately and also had short ranges because of the low-powered pistol rounds they employed.

The Red Army and the German Army made most widespread use of the submachine-gun. The most common Soviet design (around 6 million made) was the 7.62mm PPSh-41. It was cheap but robust and its drum magazine held a very useful 71 rounds. Germany produced a series of weapons based on the 9mm (0.345in) MP38 and MP40 (which were erroneously referred to as the Schmeisser) that were effective and saw widespread service.

The principal design used by both the British and American armies in the early years of the war was the 0.45in (11.4mm) Thompson, more accurate and reliable than some, but heavy and expensive. Both Britain and

Above: Soviet infantry attacking. The man nearest the camera has a PPSh-41 submachine-gun. The men on the left have Mosin-Nagant M1891 7.62mm rifles, an old design with only a five-round magazine.

the USA also produced utility wartime designs: the 0.45in M3 "Grease Gun" for the USA and the 9mm Sten for Britain. British Stens were particularly inaccurate and of dubious reliability, but they were very cheap and easy to make – an important consideration when equipping a mass army from scratch after Dunkirk. Many were also sent to resistance groups.

ASSAULT RIFLES

In the later years of the war, the Germans introduced a number of self-loading assault rifles that included many of the virtues and avoided some of the vices of both the traditional rifle and the submachine-gun. The most important design was the Sturmgewehr 44, which used a new 7.92mm (0.312in) round that would be employed in the post-war years by early models of the AK-47 Kalashnikov.

STEN GUN

Introduced in 1941, the Sten was cheap and quick to make. Over 4 million were produced in several slightly different marks. As well as being used by the British Army, it was sent to resistance forces who liked its simplicity and that it could use German ammunition.

SPECIFICATION: Sten Mark 2
CALIBRE: 9mm (0.354in)
LENGTH: 762mm (30in)
WEIGHT: 2.96kg (6lb 8oz)
BARREL LENGTH: 196mm (7.75in)
MUZZLE VELOCITY: 381m/sec (1,250ft/sec)
MAGAZINE: 32 rounds

In addition to the above, many combatants and non-combatant personnel in land, sea and air forces of all nations carried pistols (both revolvers and automatics) in a very wide variety of designs and calibres. These were used in action often enough as close-range weapons of last resort. However, they were seldom regarded as first-choice combat weapons.

Soviet Winter Victories, 1942–3

Inspired by their success around Stalingrad, Soviet attacks continued through the winter of 1942–3. Now stronger and better-equipped than ever before, they regained all the territory that had been lost during 1942.

The encirclement of Stalingrad and the eventual destruction of the German Sixth Army was only part of a more general Soviet offensive all along the southern Eastern Front in the winter of 1942–3.

General Manstein's attempt to relieve Stalingrad began on 12 December 1942 and made good progress at first, despite the bitter winter weather. Within days, however, new Soviet attacks on his flanks were posing an additional threat. The Italian Eighth Army to the north

Left: General Konstantin Rokossovsky commanded the Don Front around Stalingrad in 1942–3.

was smashed, even as the forces moving toward Stalingrad were being slowed by fierce resistance. By the end of the year Manstein's units were in retreat. To the south the Soviet Stalingrad Front (or Army Group) was striking toward Rostov to cut the German Army Group A off in the Caucasus; the Don Front was tightening the ring around Stalingrad; and the South-West Front was freeing the whole area to the west of the great Don River bend.

NEW SOVIET ATTACKS

In January the Soviet attacks were extended to the north. Once again troops from one of Germany's satellites were the initial target. Voronezh Front smashed into Hungarian Second Army, as weak and ill-equipped as its Italian and Romanian counterparts. Within days the Hungarian force had been effectively destroyed and German Second Army to its north had also been pushed back.

Off to the south Hitler had delayed giving Army Group A permission to retreat from the Caucasus. The motorized units of First Panzer Army managed

WINTER BATTLES, 1942–3
Despite their brief recovery at Kharkov, the winter battles were disastrous for the German forces.

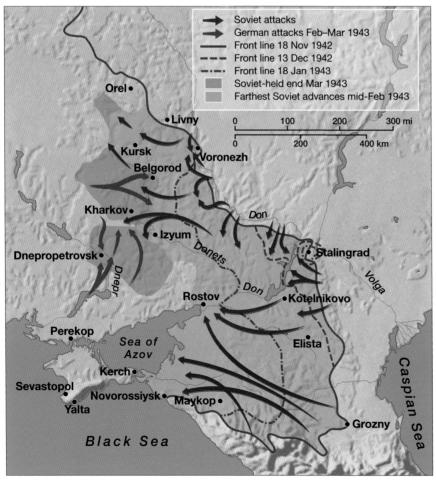

Soviet attacks
German attacks Feb–Mar 1943
Front line 18 Nov 1942
Front line 13 Dec 1942
Front line 18 Jan 1943
Soviet-held end Mar 1943
Farthest Soviet advances mid-Feb 1943

Orel
Livny
Kursk
Belgorod
Voronezh
Kharkov
Izyum
Don
Dnepropetrovsk
Denets
Dnepr
Stalingrad
Rostov
Don
Kotelnikovo
Volga
Perekop
Sea of Azov
Elista
Kerch
Sevastopol
Novorossiysk
Maykop
Yalta
Grozny
Caspian Sea
Black Sea

0   100   200   300 mi
0   200   400 km

ROMANIANS ON THE EASTERN FRONT

Romanian troops played a small part in Operation Barbarossa in 1941. The head of government, Marshal Ion Antonescu, saw this as part of a holy crusade against Bolshevism and agreed to increase the Romanian force in 1942. When the Soviets smashed the Axis defences around Stalingrad, over 100,000 Romanians were killed or captured. Even so, strong Romanian forces would continue to fight alongside the Germans until August 1944, when their country changed sides after a coup led by King Michael deposed Antonescu. Substantial Romanian forces then advanced with the Red Army into Hungary and Czechoslovakia.

Above: Romanian prisoners near Stalingrad, late 1942.

to escape through Rostov by the end of January, but the slower-moving infantry of Seventeenth Army were pushed east into a bridgehead in the Taman Peninsula opposite the Crimea.

The Soviet commanders now had new objectives in their sights. On the northern flank they would push forward to both Kursk and Kharkov, and to the south they would cross the Donets and head for the Dniepr to isolate Manstein's forces. These attacks began in the last days of January.

The Soviets also planned to strengthen these attacks by switching the troops that had finally captured Stalingrad to the main battlefront. However, this move took longer than was originally planned.

LAST ADVANCES

Soviet advances did capture Kursk and Rostov in early February and were moving past Kharkov by the middle of the month, but the troops were becoming increasingly worn out. In mid-February Hitler re-organized his forces, putting Manstein, his most able general of all, in charge of what would now be called Army Group South. The great Russian advances had created a number of exposed salients at the same time, because the German retreat had shortened their lines. All this added up to an opportunity for a German riposte, which would begin in mid-February.

Below: Red Army infantry attack. Riflemen advance while a light machine-gun gives covering fire.

Below: Soviet troops often went into action riding on tanks, a practice that risked high casualties.

Assault Guns

*Tank turrets were expensive and difficult to build so heavily armoured vehicles,
carrying powerful guns in simpler limited-traverse mountings, were also produced
in numbers and used very effectively, notably on the Eastern Front.*

Weapons in this category were almost exclusively the preserve of the German and Soviet armies. They had at least a reasonable degree of armour protection and carried weapons suitable both for anti-armour use and for direct fire support of assaulting troops and tanks. (Self-propelled guns in Anglo-American service are described in the self-propelled artillery and anti-tank gun categories – along with other Soviet and German designs – which better describe their capabilities and operational uses.)

Even with these limitations the number of vehicles that belong in this category is quite large. Soviet types include the SU-45, -57, -76, -85, -100, -122 and -152, and JSU-122 and -152 designs, too (the figures indicate the calibre of gun fitted). Germany fielded a similar variety. Accordingly, only a few examples on each side can be described.

SU-85

The SU-85 was designed to provide better anti-tank performance than the T-34/76. It was produced during 1943–4 but phased out when the T-34/85 entered service. The example shown is in German service after being captured.

WEIGHT: 29.4 tonnes
HULL LENGTH: 5.92m (19ft 5in)
HEIGHT: 2.54m (8ft 4in)
ARMAMENT: 85mm (3.35in)
  D-5 M1943
ARMOUR: 54mm (2.13in) max.
ROAD SPEED: 55kph (34mph)

FIRST DESIGNS

Germany's Sturmgeschütz 3, based on the Panzer 3 chassis, was the first notable weapon of this type. Versions of this design would serve throughout the war from 1940 and, indeed, become Germany's most-produced armoured fighting vehicle (AFV), with over 9,000 made.

Initially the StuG 3 was designed solely for infantry support with a short (L/24 – 24-calibre) 7.5cm (2.95in) gun mounted in the forward superstructure with limited traverse. Later models added first an L/43 7.5cm and then a more powerful L/48 gun to gain an effective anti-tank capability. There was also an essentially similar StuG 4 and a version fitting the 10.5cm (4.13in) howitzer. A further vehicle with comparable capabilities was the Hetzer, a design based on the PzKpfw 38(*t*) chassis and also carrying the L/48 7.5cm gun.

In a different category were various *Jagdpanzer* (usually translated as "tank destroyer") vehicles (confusingly, the Jagdpanzer 4 was really an updated StuG 4, rather than an entirely new type). Three vehicles of this sort should be mentioned.

First was the Elefant, or Ferdinand, based on an alternative design for the Tiger tank. It carried the most formidable

Left: A knocked-out Jagdpanther being examined by an American soldier in early 1945.

Above: A StuG 3 passes by a knocked-out T-34 during fighting in Poland in 1944.

SU-100

Like the SU-85, the SU-100 was based on the T-34 chassis, but with the more powerful 100mm D-10 gun. Full-scale production started in September 1944 and it saw significant service in 1945.

WEIGHT: 32.5 tonnes
HULL LENGTH: 5.92m (19ft 5in)
HEIGHT: 2.54m (8ft 4in)
ARMAMENT: 100mm (3.94in) D-10 M1944
ARMOUR: 75mm (2.95in) hull front
ROAD SPEED: 48kph (30mph)

version of the famous "eighty-eight", the PaK 43 L/71 8.8cm (3.46in) gun, behind very thick armour. Mechanically unreliable and lacking secondary armament to fend off infantry attack, it was not a success.

The Jagdpanther carried the same gun and was well armoured and agile; overall it was probably the most effective tank destroyer of the war. Its larger stablemate, the Jagdtiger, mounted a massive 12.8cm (5.04in) gun, the most powerful anti-tank gun of the war, behind armour up to 250mm (9.84in) thick, but was clumsy and unreliable. Most tellingly, for all their power, the total production of these types was roughly 90 Elefants, 390 Jagdpanthers and 80 Jagdtigers – never enough to stave off Germany's defeat.

SOVIET RESPONSES

The first significant Soviet design was the SU-76, which carried the M1942 76.2mm (3in) gun in an open-topped mount on a light-tank chassis. It was designed in response to the German StuG types; over 12,000 were

JAGDTIGER

The Jagdtiger first saw service in late 1944 and a few fought in the Battle of the Bulge. The basic vehicle was a variant of the Tiger 2 chassis, but it was under-powered and hence prone to breakdowns. Only two battalions using Jagdtigers were formed.

WEIGHT: 70 tonnes
HULL LENGTH: 7.39m (24ft 3in)
HEIGHT: 2.95m (9ft 8in)
ARMAMENT: 12.8cm (5.04in) PaK 44 + 1 x machine-gun
ARMOUR: 250mm (9.84in) max.
ROAD SPEED: 38kph (24mph)

built. Substantially more powerful were the SU-85 introduced in 1943 and SU-100 of late 1944. Neither of these had the same scale of armour as the German tank destroyers, but the SU-100 in particular had the gun power to deal with most German AFVs. By way of comparison at least 1,500 SU-100s were built by mid-1945.

The heaviest Soviet assault-gun types carried 122mm (4.8in) and 152mm (5.98in) guns. These weapons were not primarily designed for anti-tank use, but their heavy shells meant that they had significant anti-tank capabilities nonetheless and were often used in this role.

Propaganda, Art and Popular Culture

Since WWII was indeed a total war, writers, artists, movie-makers, journalists and broadcasters everywhere played their parts in the war effort, either within the dictates of totalitarian rulers or under the gentler controls imposed in the democracies.

In the totalitarian nations, art, entertainment and public information were simply aspects of national life to be managed for the good of the state, whether in war or in peace. The situation was less clear-cut in the democracies, but generally speaking popular and more serious artists and those in the media felt an obligation to make some contribution to their respective war efforts.

ABSOLUTE CONTROL

There were many similarities between the situation in the USSR and Germany and even Japan. In all three countries censorship was absolute; nothing was published or broadcast

Below: *Signal* magazine was issued by the Germans in several languages. This French example publicized the Katyń massacre.

that had not been officially approved, nor was any criticism of the ruling regime tolerated. The Nazis were particularly aware of the power of radio broadcasting. They ensured that cheap radio sets were widely available in Germany but confiscated many radios in occupied countries, banning anyone under their control from listening to Allied broadcasts.

In the visual arts there were also many similarities between the socialist realism style demanded by Soviet authorities and the classicism favoured by Hitler and the Nazis. The stern Aryan soldiers of German war art were little different from the undaunted workers and peasants their opponents depicted.

German and Soviet portrayals of their opponents were nasty in the extreme. The Nazi demonization of communists and Jews needs no elaboration, while on the Soviet side well-known and much-publicized material included poems entitled "Kill Him" and "I Hate".

Depictions of the Axis powers in Britain and the USA were neither as crude nor as vicious, though some well-known wartime films, *The Life and Death of Colonel Blimp* for example, were criticized in some quarters for including characters who were decent Germans. One thing that almost all belligerent nations on both sides had in common was that newspaper circulations increased substantially, though

JOSEF GOEBBELS

Goebbels (1897–1945) was the Nazi Propaganda Minister from Hitler's accession to power in 1933 until, after murdering their six children, he and his wife committed suicide in Hitler's bunker in April 1945. He controlled all the media and artistic output in Germany. He used radio broadcasting particularly brilliantly to foster support for the Nazi regime and to create and maintain the image of Hitler as a great leader. He was loyal to Hitler to the end and his efforts played a large part in keeping the German people fighting effectively long after defeat was certain.

Above: Josef Goebbels in full oratorical flow at a Nazi Party rally in 1934.

Above: "Behind the enemy powers: the Jew", a German poster of 1942.

in most countries the number of different publications decreased, while the survivors were reduced by paper rationing to a handful of pages.

BROADCAST MEDIA

In Britain and the USA, owners and managers of the press and broadcast media were happy generally to comply with official wishes that they gave no secrets to the enemy and to moderate any criticisms of the government that they might have. But neither government sought to dictate what was printed or broadcast and certainly did not disseminate the sort of outright lies that were a commonplace of Nazi bulletins.

Many in Europe secretly listened to the BBC, in part because they knew that its news was truthful as far as it went, even if it was not always the whole truth. But Germany's English-language broadcasts

Right: Laurence Olivier as the king in his 1944 film, *Henry V*.

were mocked in Britain for the stilted style of the principal broadcaster William Joyce (known as "Lord Haw-Haw") and the ludicrous stories he often tried to tell his listeners.

Film was probably the most powerful cultural medium. As well as productions on topical subjects, the output both from Hollywood and the state-run German film industry included much simple escapist entertainment for difficult times. Historical epics featuring past patriots and heroes appeared regularly in Germany and the USSR – *Bismarck* and *Ivan the Terrible* among the titles – and in Britain Laurence Olivier's 1944 version of Shakespeare's *Henry V* had a similar theme.

For many, popular music and song were a source of comfort and pleasure that could transcend boundaries. At the end of the African campaign in 1943 there was an incident in which a

Above: Underground newspapers, like France's communist *l'Humanité*, provided an alternative source of news in every occupied country.

formation of British soldiers, marching to their victory parade in Tunis, passed a column from the *Afrika Korps* going into captivity – both groups were singing "Lili Marlene".

Alamein and the Advance to Tunisia

Although it could not compare in scale to the massive battles on the Eastern Front,
El Alamein is rightly seen as a turning point in Britain's war with Germany.
Much hard fighting remained but after Alamein there would be no more big defeats.

Although his charge into Egypt had been halted in the First Battle of El Alamein in July 1942, Rommel was keen to attack again as soon as his forces were strengthened. In the meantime, Prime Minister Churchill had appointed a new command team to the British forces defending Egypt, putting General Bernard Montgomery in charge of Eighth Army.

NEW BRITISH PLANS

As soon as he took over, "Monty" began transforming the British force, beginning by rebuilding its morale, which had been shattered by the disasters earlier in the year. He laid down far more clearly than before that there would be no retreat and deployed his forces to make best use of their concentrated firepower, rather than try to

NORTH AFRICA, 1942–3 After victory at El Alamein and the Torch landings, the Allied forces steadily completed their African victory.

**KEY FACTS –
EL ALAMEIN**

PLACE: North-west Egypt

DATE: 23 October–4 November 1942

OUTCOME: Italian and German forces defeated and forced to retreat to western Libya.

Above: A Crusader Mark 3 tank during the Battle of El Alamein. Some 4,500 of all versions of the Crusader were built, but all were mechanically unreliable.

---

compete with the Germans in manoeuvre. Thus, when the Germans attacked late on 30 August, the British units stood firm and by 6 September the Germans had been forced back to where they had started from. This Battle of Alam Halfa was a clear Allied victory.

A significant factor in the German defeat was their short-age of supplies. They were operating many hundreds of kilometres from their nearest port, while the routes between Europe and North Africa were coming under increasingly heavy Allied air and naval attacks. Many of these raids were coming from the island of Malta, very heavily bombarded earlier in the year but by now reinforced after a series of fierce naval convoy battles.

After Alam Halfa, Churchill wanted Montgomery to attack straight away but he refused, insisting that he be allowed to build up his forces substantially first. By the time he did attack

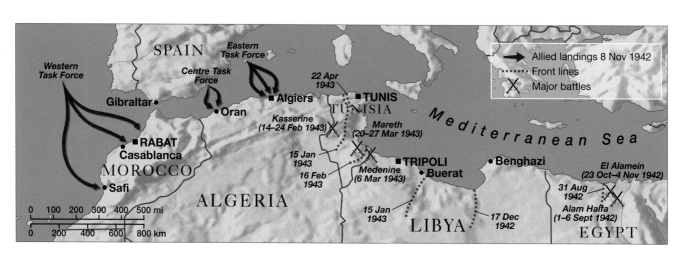

Right: A British 5.5in (140mm) gun firing during the heavy bombardment that opened the El Alamein battle.

at El Alamein in October, Eighth Army had about a two to one superiority in men, tanks and guns and a considerable advantage in the air (and much of the Axis force was made up of less effective Italian units).

Rommel had built up a formidable position by using extensive minefields and other defences, but Montgomery had trained his troops meticulously. He made certain that his infantry, tank units and artillery worked together effectively, which had seldom happened in the previous desert battles.

ALAMEIN VICTORY

Several days of attritional fighting followed the initial Allied advance. Gradually, the German tank units, which provided the backbone of the Axis defences, were worn away. On 2 November Rommel signalled to Hitler that he had to retreat. Hitler responded with his usual unrealistic "no retreat" order, but it made no difference. With only a couple of dozen of the 500 tanks with which he began the battle left in action, Rommel

Below: Italian troops run for cover under Allied air attack in the early stages of the Battle of El Alamein.

had no alternative. On 4 November he abandoned his whole defensive position.

Partly because Montgomery did not organize an energetic enough pursuit, much of the remaining Axis force got away. A long retreat followed, made even more necessary for Rommel this time because of the Anglo-American Torch landings in north-west Africa from 8 November.

After brief pauses to allow supplies to catch up, Eighth Army captured Tripoli on 23 January 1943, after an advance of some 1,500km (930 miles). Despite German demolitions, the British were able to begin using the port facilities to some extent by the end of the month.

It would be several weeks before the leading troops could be strengthened for further substantial advances, but they did manage to push forward to Medenine, in eastern Tunisia, in early February.

FIELD MARSHAL BERNARD MONTGOMERY

Montgomery (1887–1976) was Britain's most famous general of the war. His careful planning and methodical leadership played a large part in the succession of Allied victories in North Africa, Italy and north-west Europe from autumn 1942 to the end of the war. In particular, it is unlikely that D-Day would have gone so well without his contribution to the planning process. However, some historians believe that he was over-cautious and thereby lost opportunities for cheaper and quicker victories.

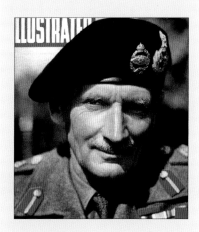

Above: "Monty" showing off his trademark pair of cap badges.

Light Armoured Vehicles

With the increasing mechanization of warfare generally, it was natural that the scouting and other support roles were filled by armoured vehicles. Armoured cars and personnel carriers accordingly proliferated in all armies in the European war.

Every army in World War II used armoured vehicles in scouting and support roles. Dozens of types were produced and most of these had numerous substantially different variants, so naturally only a selection can be discussed here.

BRITISH AND US TYPES

It is appropriate to begin with the most-produced armoured vehicle ever, the British Army's Bren, or Universal Carrier, of which over 100,000 were made. It could carry a machine-gun or mortar and its ammunition into action, tow a light anti-tank gun or serve in the scouting role, among many other tasks.

Britain was also a substantial user of wheeled armoured cars and their unarmed cousins – the scout cars. The most-produced types were the Daimler Dingo and Humber scout cars. Over 6,000 of the two-man Dingo were made. It had 30mm (1.2in) armour and such refinements as

Above: A British Humber armoured car in northern France in 1944.

run-flat tyres and transmission with five forward and reverse gears. Humber and Daimler (in that order) made the most common armoured cars too, both types being in use from 1941. Heavy armoured cars included the indigenous AEC design and the American-built T17 Staghound. Britain also relied on the USA for various (usually half-tracked) personnel carriers and similar vehicles.

Unlike Britain, the USA built and used relatively few armoured car types. By far the most important was the M8, known as the Greyhound in British service. This carried a 37mm (1.46in) gun and had

Left: An American M8 armoured car during training for D-Day "somewhere in England" in 1944.

originally been conceived for the tank-destroyer role. The most important American scout car was the White M3, with over 20,000 built. This could carry up to seven men plus the driver.

The US and German Armies made extensive use of half-tracked vehicles. These had much of the cross-country performance of fully-tracked vehicles but were easier and cheaper to build because of their

SDKFZ 251

Initially designed to carry the infantry in armoured divisions, the SdKfz 251 appeared in over 20 variants for other roles, including anti-aircraft, anti-tank and command vehicles. About 13,500 were built in all.

WEIGHT: 7.9 tonnes
LENGTH: 5.8m (19ft)
HEIGHT: 1.75m (5ft 9in)
WIDTH: 2.1m (6ft 11in)
ROAD SPEED: 53kph (33mph)
ENGINE: 120bhp Maybach HL42
ARMOUR: 15mm (0.6in) max.

Above: An M3 half track leads a US column in Germany, 1945.

simpler wheeled steering. US models included the smaller M2 and M9, and the larger M3 and M5. The M3 and M5 were 9-tonne vehicles that could carry a full infantry squad. All four were widely used in general front-line transport roles. In addition there were many variants carrying anti-aircraft, anti-tank and close-support weapons. Many US half tracks and scout cars were supplied to the Soviets.

GERMAN DESIGNS

The smallest and most unusual vehicle in Germany's half-track class was the SdKfz 2 Ketten-krad, in effect a half-track cargo motorcycle, of which over 8,000 were made. More important were the more conventional 5.9-tonne SdKfz 250 and its derivatives, as well as the larger 7.9-tonne SdKfz 251 and its numerous variants. These types were a little less mobile than the

US M2/M3, in part because the front wheels were unpowered, but they served in a similar variety of personnel, transport and weapon-carrying roles. Many were used as command vehicles, being fitted in the early-war years with large and conspicuous "bedstead" aerials for the radios they carried.

Germany also had a variety of 4-, 6- and 8-wheeled armoured cars. The 4-wheel SdKfz 221 was a 4-tonne vehicle with two crew and armed with a single machine-gun. The 6-wheel types were pre-war designs and were mostly withdrawn from service by 1941 or so. They were superseded by the 8 x 8 types from 1937.

The 8-wheel SdKfz 232 carried a 2cm (0.79in) cannon and a machine-gun (confusingly, there was also a 6-wheel SdKfz 232 and other overlapping designa-

Right: The British AEC Mark 3 heavy armoured car carried a 75mm (2.95in) gun.

BREN GUN CARRIER

The Universal Carrier (as the standardized examples were known from 1940) was the principal British utility vehicle. It often carried light weapons like the Boys anti-tank rifle and Bren Gun, seen here.

SPECIFICATION: Universal Carrier
WEIGHT: 4.3 tonnes
LENGTH: 3.75m (12ft 4in)
HEIGHT: 1.6m (5ft 3in)
WIDTH: 2.1m (6ft 11in)
ROAD SPEED: 32kph (20mph)
ENGINE: 85bhp Ford V8
ARMOUR: 10mm (0.4in) max.

tions). It was an 8.8-tonne vehicle with a crew of four, including a second driver in the rear of the fighting compartment who could drive the vehicle in reverse. The heaviest variant was the SdKfz 234 Puma, with a 7.5cm (2.95in) anti-tank gun.

Operation Torch and Victory in Africa

American troops made their combat debut in the European theatre in Operation Torch in 1942. After setbacks and much hard fighting, the Allied victory in North Africa was completed in May 1943, as the ring around Germany began to tighten.

From early 1941 US policy in the event of war with Germany and Japan was to give the European theatre priority. From the start US Army leaders were convinced that the way to defeat Hitler was by a cross-Channel attack from England, but it was very soon obvious that this operation could not be mounted for many months. To maintain the "Germany first" policy President Roosevelt ordered that US troops must be sent into battle against Germany during 1942, so it was decided to make landings in Morocco and Algeria to clear North Africa of Axis forces.

Morocco and Algeria were French colonies under the control of Marshal Pétain's Vichy

Below: An American M2 105mm (4.13in) howitzer position in southern Tunisia in the spring of 1943.

KEY FACTS – TORCH

PLACE: North-west Africa

DATE: From 8 November 1942

OUTCOME: Allied forces landed successfully in Morocco and Algeria, but German reinforcements poured into Tunisia to slow their advance.

regime. Anglo-French relations remained embittered after the events of 1940 so the Americans took the lead in pre-landing negotiations to try to persuade the leaders of the large Vichy forces in Africa not to resist the Allied advance and, instead, join with the smaller Free French forces already fighting on the Allied side. In the event the French did not resist the land-

ings strongly, but it took months of political wrangling to create a united Free French force.

TORCH LANDINGS

The invasion, known as Operation Torch, began on 8 November 1942, four days after Eighth Army's victory at El Alamein had sent the Italians and Germans reeling out of Egypt. Ideally the attack would have gone as far east as Tunisia, but this was rejected as too risky. Instead landings were made on Morocco's Atlantic coast near Casablanca and around Algiers and Oran in Algeria. The Anglo-American force, First Army, immediately began heading east for Tunis, but too little too late.

Field Marshal Albrecht Kesselring, Germany's very able Commander-in-Chief South, started pouring troops, tanks and aircraft into Tunisia from Sicily on 9 November. These fought the Allied advance to a standstill in the rugged hills of western Tunisia, in a series of bitter battles from late November into early January 1943.

BATTLE FOR TUNISIA

By February 1943 the Axis forces in Tunisia had been joined by Rommel's army, which had successfully completed its retreat from Egypt. Eighth Army was still struggling to bring the port of Tripoli into full use so was not yet ready to push farther into Tunisia in the south. This gave

Above: American troops landing at Oran with a light anti-aircraft gun at the start of Operation Torch.

In April First and Eighth Armies continued their attacks. They eventually broke out of the hills and advanced to Bizerta and Tunis in early May. The last of around 250,000 Axis troops taken prisoner in these final battles surrendered on 13 May. Africa was now wholly in Allied hands.

the Germans an opportunity to strike at First Army, which they took on 14 February.

In the resulting Battle of Kasserine the inexperienced American troops, who were the initial target for the attack, were badly defeated. American and British reinforcements, rushed to the threatened sector, just prevented a decisive breakthrough and then gradually recovered much of the lost ground by late February. General George Patton took over command of the US ground forces in Tunisia and his leadership very quickly improved their combat efficiency.

In early March Eighth Army smashed an attempted German attack and later in the month, in the hard-fought Battle of Mareth, pushed the Axis forces out of their main defence line in southern Tunisia.

Below: US troops examining an abandoned Italian M13/40 tank during the fighting in Tunisia.

GENERAL DWIGHT D. EISENHOWER

Eisenhower (1890–1969) was appointed as Allied supreme commander for Operation Torch and would continue in that role in Sicily and Italy and then finally in north-west Europe in 1944–5. He realized far better than other top Allied commanders the paramount importance of maintaining good relations within the Anglo-American partnership. Although he had numerous problems with egotistical subordinates like Patton and Montgomery, Eisenhower's leadership kept the Allied effort very much on track.

Above: Eisenhower was totally committed to Allied unity.

Germany Fights Back: Kharkov and Kursk

The counter-stroke at Kharkov inspired Hitler to order new attacks at Kursk. For the first time a major German summer offensive in Russia failed to break through, a clear sign that the Soviets were now winning their Great Patriotic War.

The Soviet advances during January and February 1943 had pushed deep into German-held territory to capture the city of Kharkov and threaten Dnepropetrovsk. To the north of this penetration the German Army Group Centre was holding firm around Orel, and to the south General Manstein's Army Group South had good defensive positions along the Mius River. Manstein now had several Panzer divisions available, including strong and well-equipped Waffen-SS units. He was ready to turn the tables on the Soviet attackers in between these bastions.

KEY FACTS – KURSK

PLACE: Between Orel and Kharkov, Eastern Front

DATE: 4–13 July 1943

OUTCOME: German forces failed to break through in the biggest tank battle of the war.

Below: German troops fighting on the outskirts of Kharkov at the end of February 1943.

BATTLE OF KHARKOV

In the second half of February Manstein's tank units struck against the Soviet spearheads near Dnepropetrovsk. By the end of the month four Soviet tank and mechanized corps (each roughly equivalent to a German Panzer division) had been smashed and the survivors pushed back over the Donets.

Taking advantage of the last few days of firm frozen ground before the mud of the spring thaw, Manstein's tanks next headed for Kharkov, taking it by 16 March. Hitler was so heartened by this remarkable comeback that, even before Kharkov fell, he was issuing orders for a new attack against the large Soviet salient centred around Kursk to the north.

OPERATION CITADEL

Although the Germans had lost perhaps half a million casualties over the winter, Hitler still believed he could regain the initiative on the Eastern Front and, at the same time, make provision in western and southern Europe for whatever the Anglo-Americans might attempt during the year. It was clear that the German troops, and especially their tank forces, were still usually more skilled in combat than the Red Army. Hitler also believed that new Tiger and

Panther tanks coming into service would give them a further qualitative edge.

The German plan was for Ninth Army from Field Marshal Günther von Kluge's Army Group Centre to attack the north flank of the salient, while Fourth Panzer Army from Manstein's Army Group South drove into the southern flank. Originally Hitler planned the advance to begin in May, but he repeatedly postponed it until more of the new tanks were ready.

For their part Stalin, Zhukov and Chief of the General Staff Marshal Aleksandr Vasilevsky (in effect the command team who would run the remainder of the Soviet Union's war) decided to stand on the defensive at first, while they also prepared offensives of their own north and south of Kursk. Through spies and other intelligence the Soviets had a reasonable idea of what the Germans planned and accordingly they massively re-inforced their defences in the threatened areas.

The German advances began on 4 July and were fiercely resisted from the start. In the north

Ninth Army, with seven Panzer divisions leading its attack, made a little progress at first. The Soviet commanders then committed their tank reserves. On 9 July Kluge told Hitler that he could not break through. On the 12th the Soviets began their offensive north of Orel and Kluge had to retreat.

On the southern front things seemed to go better for the Germans. Fourth Panzer Army advanced about 35km (20 miles) in the first few days, losing heavily but inflicting more casualties than it received. On the 12th there was a giant tank

Above: German Panzer 4 tanks advancing across open ground during the Battle of Kursk.

battle near the small town of Prokhorovka. Again, though the Soviets probably came off worse, they had outnumbered the Germans from the start and could afford to do so.

On the 13 July, with no breakthrough in sight and the Anglo-American invasion of Sicily developing, Hitler called off further attacks. With that, the initiative on the Eastern Front finally and permanently passed to the Red Army.

FIELD MARSHAL ERICH VON MANSTEIN

The most talented general of any country during WWII was Germany's Field Marshal von Manstein (1887–1973). He was very largely responsible for the brilliantly successful German plan for attacking France in 1940. Although he failed in the attempt to relieve Stalingrad in late 1942, his leadership over the following months stabilized the Eastern Front and led to the German success at Kharkov. At Kursk his attacks were the most successful part of the German offensive. Thereafter, Manstein fought resourcefully on the retreat until Hitler dismissed him in March 1944.

Right: Field Marshal von Manstein and his staff studying a map to plan their manoeuvres.

Ground-attack Aircraft

Attack operations close to the front line were probably the most effective uses made of air power during the war. The Stuka symbolized Germany's early-war success and the Allied riposte was led by the Shturmovik, Typhoon and Thunderbolt.

In 1939 only the German and Soviet air forces laid any stress on the ground-attack mission. Germany's close integration of land and air power in the Blitzkrieg campaigns of 1939–41 proved that such operations could be very effective indeed. Britain and the USA would develop this capability in the course of the war.

Germany's Junkers 87 Stuka dive-bomber in effect became the definitive image of Blitzkrieg and was very much feared by opposing forces. In fact it was slow and poorly protected, as was shown when it first faced serious opposition during the Battle of Britain. However, it continued to serve successfully in the early Eastern Front battles – and the Ju-87G model, available from 1943, was fitted with a pair of 3.7cm (1.46in) cannon for the tank-busting role. It served very effectively in this mission, most notably in the hands of Hans-Ulrich Rudel, an ace pilot credited with destroying over 500 Soviet tanks.

Other German ground-attack types included versions of the Focke-Wulf (Fw) 190 fighter and the Fw 189, which was also used for reconnaissance. The Henschel 129 also served in small numbers, with armament including a 7.5cm (2.95in) gun.

THE RED AIR FORCE

By 1945 the Soviets had the world's most powerful tactical-support air force, created out of the ruins left by Germany's onslaught in 1941. Although the Soviet leaders had decided to concentrate on tactical aviation shortly before the war, their aircraft designs and training had not caught up with this change when Operation Barbarossa began.

There were various obsolete fighters in service in the attack role, including the biplane

Below: Ground crew bombing up a Typhoon fighter-bomber. It carries the "D-Day stripes" used in 1944–5.

P-47D THUNDERBOLT

The Republic P-47D Thunderbolt was particularly large and heavy for a WWII fighter, but its ample power meant that it could carry a heavy attack load. P-47D pilots claimed the destruction of tens of thousands of German tanks and trucks in Europe in 1944–5.

CREW: 1
ENGINE: Pratt & Whitney R2800 radial, 2,535hp
SPEED: 697kph (433mph)
ARMAMENT: 8 x 0.5in (12.7mm) machine-guns + 10 x 5in (127mm) rockets and/or up to 1,130kg (2,500lb) bombs

Above: A Ju-87G Stuka armed with 3.7cm (1.46in) anti-tank cannon.

Ilyushin (Il) 153 and the more modern Sukhoi 2. Just entering service was something much better – the Il-2 Bronirovanni Shturmovik (the "Armoured Attacker"), which eventually would become the most-produced military aircraft ever.

The Shturmovik was in effect a flying tank, with substantial armour protection for the crew compartment and other vital parts. It had powerful cannon and machine-gun armament (varying between models), backed by a substantial load of rockets and bombs. Formations of Shturmoviks would often circle round a German position or tank unit, making repeated attacks until their target had been smashed, a tactic called the "Circle of Death". In Soviet eyes it was the most important aircraft of the war.

Secondary, but still produced in substantial numbers (over 11,000), was the Petlyakov 2. This twin-engined design had

been conceived as a high-altitude fighter but was converted to the attack role. It was robust and fast and could carry 3 tonnes of bombs.

THE WESTERN ALLIES

In the absence of appropriate aircraft, Anglo-French ground-attack operations in 1940 were disastrous. Subsequently Britain began using versions of the Hurricane and P-40 Kittyhawk fighters in the role. These could carry a useful bomb load and had effective cannon and machine-gun armament – that included a pair of 40mm (1.58in) guns in one Hurricane variant) – but were lacking in performance when so equipped.

Fighter-bomber versions of the Bristol Beaufighter and De Havilland Mosquito were also employed, with some success. Both these large and powerful aircraft could carry formidable weapons loads but accuracy in attack remained uncertain.

From 1943 both British and US attack aircraft began using rockets. Although these lacked

IL-2M SHTURMOVIK

The prototype Shturmovik first flew in 1939 and a few were in service in 1941. Initial models were single-seaters but most production was of the two-seat Il-2M and -2M3 versions, which had various other improvements, too. Further changes led to the Il-10, which saw some action in 1945.

CREW: 2
ENGINE: Mikulin AM-38 in-line, 1,680hp
SPEED: 414kph (257mph)
ARMAMENT: 2 x 23mm (0.91in) cannon, 1 x 12.7mm (0.5in), 2 x 7.62mm (0.3in) machine-guns + 4 x RS82 or RS132 rockets and/or up to 600kg (1,320lb) bombs

the pinpoint accuracy to knock out a tank, they had the ability to swamp a larger or less well-protected target with fire.

Their best-known use was when fitted to aircraft like Britain's Hawker Typhoon and the USA's Republic P-47 Thunderbolt. Both of these were originally pure fighter designs (the Thunderbolt a successful one, the Typhoon less so), which had the power and durability to blossom in the attack role. They played a vital part in the Anglo-American victory in Europe in 1944–5.

Codes and Code Breaking

The continuous struggle to read enemy messages was an ongoing aspect of every campaign of the war. Signals intelligence contributed to many victories but was probably not the war-winning advantage that is sometimes suggested.

All the major combatants in WWII put considerable effort into reading coded enemy radio messages. Overall, Britain and the USA had the most success and made the best use of the resulting information; the USA and USSR were the most successful in keeping their own messages secure (though even now little is publicly known of Soviet efforts in this field).

CIPHER SYSTEMS

Encryption methods used in WWII included manual systems (based on printed sets of random numbers) and machines, which were more complex and theoretically more secure. Signals produced on some machines are thought never to have been broken. These included the American Sigaba type and the British Typex. Less secure machines included the German Enigma and Geheimschreiber, the Japanese diplomatic service machine known to the Allies as Purple, and the American M-209 (used for low-grade traffic). Efforts to decode messages led to the development in Britain and the USA of various forms of calculating equipment, including what are now regarded as the first electronic computers.

Britain had various successes, increasingly comprehensive as the war went on, against German and Italian systems. For their part the Axis powers were able to read many Royal Navy messages up to 1943.

BREAKING ENIGMA

The British system built on pre-war French and Polish work. The first messages decoded were from the general Luftwaffe Enigma cipher in May 1940 and then other breaks followed. The process was by no means continuous. The main U-boat cipher was broken for much of 1941 but was impenetrable for most of 1942, both times with important

Left: An Enigma machine (bottom left) in use on General Guderian's command vehicle in 1940.

Above: Alan Turing, one of the leading mathematicians in the British code-breaking organization.

effects in the Battle of the Atlantic. One of the strengths of the British set-up was its centralization at the Government Code and Cypher School at Bletchley Park, where the various departments worked together well, enhancing each other's techniques and sharing information.

By contrast Germany, Italy and Japan all had a range of agencies involved in this work and these often competed with each other. And even when information was shared between services or with Axis partners, its

Below: *U-505* alongside the USS *Pillsbury*. This 1944 capture was based on code-breaking information.

Right: General Oshima (left), Japan's ambassador in Berlin, whose messages home were decoded by the Allies.

recipients did not necessarily trust it – the Italians warned the Germans about Allied code breaking and the Germans warned the Japanese, in both cases with little effect.

The German Navy's B-Dienst had good results with the main Royal Navy operational ciphers at various periods up to 1943 and systems used by Allied merchant ships into 1944. From late 1943 the manual systems were replaced in Royal Navy service by the highly secure Typex machine already used by the other British services.

In the Pacific the American-led Allied effort had great success first with the Japanese diplomatic cipher, then with the Japanese Navy and finally the Army systems. Curiously, among the most valuable results was the insight given into German plans. Japan's diplomats in Berlin and elsewhere sent extensive reports on new German weapons and defences, which were read by the Allies.

The breaking of the Japanese Navy's JN-25 cipher made a major contribution to the US victory at Midway that turned the tide of the Pacific War.

Later code-breaking information helped Allied commanders decide which garrisons to attack and which to bypass. However, it is not true (as has sometimes been suggested) that properly interpreted Purple and JN-25 decrypts made by the US, British or Dutch services could have given definite warning of Japan's attack on Pearl Harbor.

GARDENING

During their normal operations RAF bombers often dropped sea mines near German-controlled ports. This activity was known by the codename "gardening". This was not just done to sink enemy ships, however. Many such operations were meant to be detected by the Germans so that their local HQ would send a warning to its units. Allied cryptographers then used the predictable content of such signals to work out the code-machine settings for the day – and then they would use these settings to read more important messages.

The Defeat of Hitler's Navy, 1942–5

The struggle against Germany's U-boats continued to 1945 but was decisively won in the spring of 1943 by a combination of airpower, scientific research, code-breaking skill, industrial strength – and the bravery of the Allied seamen.

By the end of 1941 the battles between Germany's U-boats and the Allied merchant ships and escorts had already been raging bitterly for many months. Though the threat was serious, Britain was surviving reasonably well. Rationing had reduced consumption substantially and new construction, and the transfer to British control of ships from occupied Allied countries like Norway and Greece, had helped to counteract losses. However, Hitler's declaration of war on the USA in December 1941 brought new opportunities, which his U-boat force soon grabbed.

For the first six months or so of 1942 the U-boats scored many successes off the US East Coast and in the Caribbean. The main reason for this was an astounding failure on the part of American commanders to institute a system of escorted convoys and supporting air units. Convoys linking with

Above: A British heavy cruiser in service with an Arctic convoy to Russia in early 1943.

those on the trans-Atlantic routes were gradually introduced from April but did not cover the whole area to the Gulf of Mexico until October. Although the situation was thus being brought under control by the summer, the Allied shipping losses of June 1942 were the worst of the war.

GERMAN ADVANTAGES

On the German side matters were helped by code changes that shut the British cryptographers out of the main U-boat traffic from February 1942 to the end of the year. The size of the U-boat force was also increasing from some 100 operational boats in January 1942 to

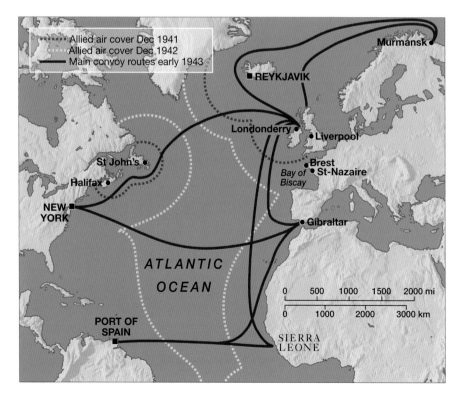

Allied air cover Dec 1941
Allied air cover Dec 1942
Main convoy routes early 1943

Murmansk
REYKJAVIK
Londonderry • Liverpool
Brest
Bay of • St-Nazaire
Biscay
St John's
Halifax
NEW YORK
Gibraltar
ATLANTIC
OCEAN
PORT OF
SPAIN
SIERRA
LEONE

0 500 1000 1500 2000 mi
0 1000 2000 3000 km

BATTLE OF THE ATLANTIC
Convoy routes and air cover areas at the height of the battle.

over 200 a year later. Ultimately, though, they were fighting a losing battle in the production race. July 1942 was the first month of the war when new ships launched on the Allied side exceeded losses, and the American shipbuilding effort was still expanding rapidly.

Although the point was past when there was any likelihood of Germany winning the Battle of the Atlantic, a further crisis was to come. This was caused by poor allocation of resources on the Allied side. Although shipping losses were delaying the US build-up in Britain, and hence the subsequent invasion of north-west Europe that was supposed to be the principal Allied plan, the forces allocated to the Atlantic battle were relatively weak. This was especially true in the vital category of very long-range aircraft, with hundreds being used by the heavy bomber forces in England and the US Navy in the

Above: The merchant ship *Pennsylvania Sun* burning after an attack by *U-571* in July 1942.

Pacific but, until well into 1943, only a handful being allocated to the convoy routes.

CLIMAX OF THE BATTLE
In a series of frantic convoy battles in March 1943 the U-boats scored their last significant victories, sinking about a fifth of the ships crossing the Atlantic at that time with very little loss to themselves. Within weeks, however, the situation was transformed. In May, 41 U-boats were sunk, and Admiral Dönitz abandoned his tactic of pack attacks on convoys.

There were many reasons for this sudden change. More Allied aircraft were committed to the battle, both ship-borne in escort carriers and land-based. Allied ship and airborne radars were much improved, as were anti-submarine weapons. Tactics and training of escort ships were much more sophisticated. And, by no means least, the British code breakers were again reliably reading German messages.

The Allies would hold their lead in all these categories until the war's end, despite German efforts to turn the tide with new

weapons and tactics. Although numerous U-boats remained in action until May 1945 (and they still achieved a few scattered sinkings), by the final months most new U-boats were being sunk before they finished their first war patrols.

Above: Admiral Ernest King was the highly effective head of the US Navy 1942–5 but was slow to set up the convoy system off the US coast.

ADMIRAL KARL DÖNITZ

Dönitz (1891–1980) was the head of the German submarine service from the start of the war and Commander-in-Chief of the German Navy from January 1943. When Hitler committed suicide in 1945, Dönitz briefly became his successor. Dönitz had been a U-boat captain in WWI and commanded the U-boat force very resourcefully throughout WWII. Dönitz thought his U-boats could win the war for Germany, but too few were built to achieve this until far too late.

Escort Carriers

No escort carriers existed in 1939, but by the end of the war Britain and the USA had well over a hundred ships of this type protecting the vital Atlantic supply routes and supporting amphibious operations in the Pacific.

One of the simplest lessons of WWII was that air power was vital in every area of warfare, by sea as much as by land. The USA, Britain and Japan were the only countries to operate large "fleet carriers" with their main naval forces and were also the only countries to deploy smaller "escort carriers" for what were sometimes seen as more mundane, second-line duties (though Japan's half dozen such completed vessels achieved nothing).

CONVOY AIR COVER

As the Battle of the Atlantic developed in intensity from the summer of 1940, the British authorities soon decided their Atlantic convoys needed air protection. Since fleet carriers were too scarce and valuable for use in this role, different ships and techniques were devised. At that point the main task envisaged for such ships was to counter the German Focke-Wulf Kondor aircraft, which were attacking convoys and homing-in U-boat packs.

The first expedient, from April 1941, was to fit a single aircraft-launching catapult to a merchant ship to carry a fighter. There were four Royal Navy fighter catapult ships and 35 merchant navy catapult aircraft merchant (CAM) ships. After his mission the pilot either had to make for land or ditch his plane near the convoy and hope to be picked up. Remarkably, this perilous process sometimes worked; six German aircraft are said to have been shot down by CAM-ship aircraft, though a number of the ships were themselves sunk by U-boats.

The first true escort carrier was HMS *Audacity*, operational in June 1941. *Audacity* only sailed with three convoys before being sunk by a U-boat in December 1941, but it had already clearly proved its worth.

By then the US Navy had taken delivery of its first escort carrier and was building more,

Below: HMS *Audacity*, a converted merchant ship, carried only eight fighters for convoy air defence.

USS *LONG ISLAND*

Commissioned in June 1941, the *Long Island* was the first American escort carrier, and like HMS *Audacity* a converted merchant ship. *Long Island* was mainly used for second-line duties, in aircrew training (as seen here in May 1943) or ferrying aircraft to combat zones. It carried the first aircraft to Guadalcanal in 1942.

DISPLACEMENT: 13,500 tons
LENGTH: 150m (492ft)
BEAM: 21.2m (69.5ft)
SPEED: 16.5 knots
CREW: 970
AIRCRAFT: 20 approx.

both for its own use and for the Royal Navy. In all some 130 escort carriers were built, or converted from existing merchant-ship or auxiliary-cruiser hulls.

The Bogue and Casablanca classes were the main types. They usually operated 20–35 aircraft, often a mix of about one-third fighters and two-thirds bomber/reconnaissance

Right: The CAM-ship *Empire Spray* with a Sea Hurricane fighter on its catapult, seen in October 1941.

types. They were much slower than fleet carriers and only lightly built, but it was never intended that they should operate in areas where air or surface attack was a significant risk.

For a variety of reasons, few escort carriers came into service on the Atlantic convoy routes until 1943. However, from then until the end of the war, they protected many convoys and hunted down and sank numerous U-boats. In June 1944 the escort carrier USS *Guadalcanal* even assisted in the capture of a German submarine, U-505.

From mid-1943 there were also 19 British merchant aircraft carriers (MAC ships), merchant ships given a very basic flight deck and three or four Swordfish aircraft equipped for the anti-submarine role. These also carried normal cargoes and sailed with convoys; no convoy that included a MAC ship ever lost a vessel to U-boat attack.

PACIFIC COMBATS

Some escort carriers were used to train carrier aircrews, but many also saw extensive combat service in the Pacific. As the US counter-offensive developed, escort carrier groups were used to provide close support to the various landing forces, while the main fleet carrier groups wore down Japanese air power and guarded against interventions by the Japanese fleet.

Right: The USS *Makin Island*, a Casablanca-class ship, seen near Leyte in November 1944 early in its combat career.

At least that was the theory. At Leyte Gulf, however, it went drastically wrong. In a remarkable action the escort carrier groups successfully fought off an attack by some of the Japanese Navy's most powerful ships. One escort carrier was sunk by gunfire in this action and another by a kamikaze aircraft.

Naval Weapons and Electronics

No entirely new naval weapons were introduced in the course of WWII, but the existing armoury of guns, torpedoes and depth charges was given new accuracy and striking power by developments in control systems and detection equipment.

Guns in service at sea in WWII ranged from the 460mm (18.1in) monsters fitted to Japan's *Yamato* and its sister ships to the light anti-aircraft weapons of 20mm (0.79in) and upward, which were carried by almost every fighting vessel of all countries' navies.

As well as firing truly formidable projectiles (1,460kg/ 3,220lb for the *Yamato*), the big guns had a very considerable reach. The longest-range hit ever made by a gun on a moving target – 24km (15 miles) – was by a 15in (380mm) on HMS *Warspite* against the Italian *Giulio Cesare* in July 1940.

Below: The carrier USS *Cowpens* in 1943. SC, SG and SK radar aerials are among the equipment on view, illustrating radar's importance.

Above: Handling 929kg (2,048lb) 16in (406mm) shells in the magazine of the battleship HMS *Nelson*.

TORPEDOES

The guns used by all navies were generally comparable in performance (though poorly manufactured Italian shells were notably inaccurate). However, this was less true with the other major anti-ship weapon – the torpedo.

There were two main torpedo propulsion systems. The more common used compressed gas and oil or alcohol fuel to drive the torpedo engine. This gave the best speed/range combination but left a wake in the water behind the torpedo, which could give the target sufficient warning to dodge. The best such torpedo was the 610mm (24in) Japanese Type 93, usually known by the nickname of "Long Lance". This torpedo used compressed oxygen, rather than air, to achieve a far better performance than any other type.

Germany, and later the US Navy, also used battery-powered torpedoes. These had shorter ranges but left no wake.

Early-war torpedoes were designed to run at a fixed depth in a straight line and to detonate either by contact or underneath an enemy ship, by using a magnetic influence device. Both the Germans and Americans had numerous problems with unreliable depth-keeping for many months after they joined the war and all nations found their magnetic influence warheads rather temperamental.

Developments during the war included German torpedoes that could follow a zigzag or looping course to increase the chances of a hit, and acoustic homing torpedoes, produced by both the Germans and the Allies, used against submerged submarines or other targets.

Above: The *Graf Spee* in December 1939, showing signs of damage after the Battle of the River Plate. The ship's Seetakt radar aerial can be seen at the top of the picture.

The main anti-submarine weapons were unguided underwater bombs known as depth charges. These simply sank through the water to explode at a pre-set depth. Typically they had to detonate within 10m (33ft) of a submarine to sink it, so several were usually dropped in a "pattern" with slightly different settings. Depth charges were improved during the war by being filled with increasingly powerful explosive compounds.

Depth charges were supplemented by smaller weapons, either contact or depth fused, which could be thrown ahead of the attacking ship. The most successful were the British Hedgehog and Squid types.

ELECTRONICS

The main underwater sensor in use was sonar (officially called asdic in the Royal Navy until 1943). This used sound pulses to find the range and bearing of a target but did not determine the depth of the submarine; it was also blind in the area underneath the ship (hence the utility of forward-firing weapons).

Radar naturally played an important part in the war at sea, detecting enemy ships and aircraft and giving gunnery ranges in bad weather and at night. Germany's Seetakt type, in service in 1939, was a highly effective early-war design. Later-war Allied designs, like the British Type 271 and others, could detect a target as small as a submarine periscope.

Above: A depth charge being deployed from a US escort.

Just as important as target detection and ranging systems was equipment to translate this and other information into firing data. The US Navy in particular developed effective anti-aircraft control systems and the American torpedo data computer fitted in submarines was superior to its equivalents in use in other navies.

Below: A 20mm (0.79in) anti-aircraft gun position on the foredeck of the battleship USS *Iowa* in 1943. Two of the ship's 16in (406mm) main gun turrets can be seen behind.

Allied Grand Strategy

Britain and the USA never agreed a formal alliance treaty during WWII but, despite this and their radically different political systems and ways of life, they generally co-operated effectively to defeat Germany.

Throughout the war there were a number of summit meetings between Churchill and Roosevelt and their top advisers. Sometimes these also involved other Allied leaders like the French and Chinese. There were also three summit meetings with Stalin, and Churchill travelled separately to Moscow during 1942.

MILITARY PLANS

More important in the very close and generally amicable Anglo-American co-operation that developed was the creation of an integrated military planning system. Britain and the USA both had chiefs of staff committees of their top soldiers, sailors and airmen to oversee their national military operations. However, a new body, the Combined Chiefs of Staff,

Above: Roosevelt and Churchill seen in contemporary coverage of the Casablanca Conference.

Below: Churchill and his Chiefs of Staff in 1945. Front row from left, Air Marshal Portal, General Brooke, Churchill, Admiral Cunningham.

was created by the first inter-Allied conference after the USA joined the war: the Washington "Arcadia" conference during December 1940–January 1941.

The Combined Chiefs normally met in Washington and included the US Joint Chiefs in person and representatives of the British Chiefs permanently assigned to this duty. This body agreed many of the vital details of major military plans and allocated resources accordingly.

At a lower level all major Anglo-American operations in the European theatre, from the D-Day invasion down, were planned and staffed by fully integrated command teams. One of the most notable achievements of General Eisenhower, as Allied Supreme Commander for a number of European operations, was to ensure that disagreements along national lines were avoided as much as possible.

The summits also included meetings of the Combined Chiefs (though in this case the British Chiefs of Staff would be personally present) and here national disagreements were perhaps inevitably more common. There were two Allied conferences at Quebec, a second one at Washington, and one each at Casablanca, Cairo and Malta. Each had complicated discussions and arguments that cannot be outlined in detail here. However, certain important themes can be highlighted.

CROSS-CHANNEL ATTACK

The American leaders consistently believed that the best way to defeat Germany was the most direct: build up forces in Britain, invade across the Channel and head straight to Berlin. From bitter experience the British were all too well aware of the fighting power of the German Army. They were unwilling to risk a major defeat by a premature cross-Channel invasion. They also believed that too single-minded a concentration on this one objective would be to neglect important interests in the Mediterranean and prevent the Allies responding successfully to opportunities that might arise.

Generally speaking the British were more successful in getting their views adopted as policy in 1942–3, when their war effort was still more substantial than the USA's, than in the later period when the USA's strength was still expanding as Britain's declined. In practice this meant that the clearance of North Africa and the subsequent invasion of Sicily and southern Italy in 1942–3 were not originally welcomed by the American Chiefs and that the later removal of many resources from the Italian campaign was bitterly opposed by the British.

The summits with Stalin (in Teheran, Yalta and Potsdam, at the last of which Harry Truman and Clement Attlee replaced Roosevelt and Churchill) were more about how post-war Europe should be organized

than about agreeing military plans. Ultimately, one truth became plain: the war would end with the Red Army in control across eastern Europe and Stalin would rule it as he saw fit, whatever wartime promises he might make regarding free elections and similar matters.

Above: Stalin, Roosevelt and Churchill during their first conference at Teheran in 1943.

However, Stalin did keep his promise, made at Teheran in November 1943, to join the war against Japan shortly after the end of the war in Europe.

Right: Managing shipping, like these newly launched American Liberty ships, was a key aspect of Anglo-American co-operation.

Medium Tanks, 1942–5

Although their activities were increasingly constrained by air, artillery and infantry weapons, tanks still played a decisive role in combat, especially the intermediate-sized designs that appeared by the thousand on most European battlefields.

As tanks developed, competition between increased gun power and thicker armour naturally continued, but even more important was sheer quantity.

MASS PRODUCTION

Although Germany's PzKpfw 5 Panther was the most formidable tank discussed here, only some 6,000 were built compared to over 40,000 Soviet T-34s and 50,000 American Shermans. (There is no exact definition of the difference between a medium and a heavy tank but included here are tanks, regardless of weight, that normally served in the armoured regiments of standard armoured divisions, rather than in separate heavy tank units.)

The T-34 had set the benchmark for future designs during its first significant combats in 1941. It continued to give effective service in only slightly modified forms into 1944. By then its 76.2mm (3in) gun was insufficient to tackle the latest German types. From early 1944 a much better version – the T-34/85, carrying an 85mm (3.35in) gun – was produced. As well as sufficient gun power to destroy a Panther at normal battle ranges, this had a three-man turret. Previously the T-34 commander was also the gunner but now he could concentrate on his main role, which greatly improved combat efficiency.

The principal Anglo-American tank of the later-war years was the American M4 Sherman. When it appeared in 1942 this was broadly comparable with the contemporary T-34/76 or the later versions of the PzKpfw 4 (which would continue in service to the end of the war). Its 75mm (2.95in) main gun then had adequate power and it matched reasonable armour with excellent manoeuvrability and reliability. It was judged to be of sufficient quality that for a time US authorities halted development work on a successor to concentrate on mass production of this design (in many minor variants). Unfortunately this proved to be

Below: A cast-hull, 75mm-armed model of a Sherman in Italy in 1944.

COMET A34

The Comet A34 was the final vehicle in the British series of cruiser tanks in service throughout the war. It was a development of the Cromwell, with a much better gun based on the 17pdr and stronger, more reliable suspension.

WEIGHT: 33.2 tonnes
ARMAMENT: 1 x 77mm (3in) gun, 2 x machine-guns
ARMOUR: 101mm (4in) max.
CREW: 5
ROAD SPEED: 50kph (31mph)

Above: A Panther in action on the Eastern Front. Far too few Panthers were built to cope with the flood of Allied Shermans and T-34s.

T-34/85

Some 22,500 T-34/85 tanks were built in 1944–5. As well as having a better gun and improved internal arrangements, it retained the wide tracks and good cross-country performance of the earlier T-34/76 models.

WEIGHT: 32 tonnes
ARMAMENT: 1 x 85mm (3.35in) ZiS S53 gun, 2 x machine-guns
ARMOUR: 90mm (3.5in)
CREW: 5
ROAD SPEED: 50kph (31mph)

an unwise decision as, by 1944–5, its weaknesses had become very plain. The 75mm gun could not penetrate the frontal armour of a Panther or Tiger at all, and when hit itself the Sherman usually quickly burst into flames – a nightmare situation for any tank crewman.

BETTER SHERMANS

Improved versions did come into service by 1944–5. Some US Shermans were fitted with a more powerful 76mm (3in) gun and had better ammunition stowage, which reduced the fire problem. But the only model with truly adequate firepower was Britain's Sherman Firefly, with a version of the 17pdr (76mm) anti-tank gun. Even this had drawbacks: rate of fire was slow and Fireflies were often singled out as priority targets by German tanks.

Britain also had indigenous designs, developed from the early-war cruiser tank series. The Cromwell, in widespread use in 1944, was very fast and, unlike its predecessors, fairly reliable. However, it carried the same 75mm gun as the Sherman and had similar armour thickness. It was developed into the

SHERMAN FIREFLY

The Firefly was a conversion, not a new-build design. The most important change was the larger turret to fit the more powerful gun, but the hull machine-gun and radio operator were also omitted and the space used for ammunition.

WEIGHT: 32.5 tonnes
ARMAMENT: 1 x 17pdr (76mm) gun, 1 x machine-gun
ARMOUR: 76mm (3in)
CREW: 4
ROAD SPEED: 39kph (24mph)

Comet, in service from the autumn of 1944, with similar virtues and a version of the 17pdr gun, making it the first wholly British tank of the war with adequate firepower.

By 1944–5 the state of the art was defined by the Panther. Judged by its weight (45 tonnes) this was a heavy tank. It might be more accurately described by the post-war description "main battle tank" for its mix of thick armour, reasonable speed and, above all, substantial gun power. Its gun was an extremely potent 7.5cm KwK 42 and it had very thick and well-sloped armour. Mobility and reliability were its weaknesses, but it was a formidable opponent.

Sicily and Italy, 1943–4

The Allied invasion of Sicily in July 1943 was followed by landings in mainland Italy in September. These knocked Italy out of the war, but the German Army's continued stubborn defence meant that there would be no rapid Allied victory.

In their meeting at Casablanca at the start of 1943, the Anglo-American leaders decided to invade Sicily as soon as possible after the campaign in North Africa was over. Accordingly, Operation Husky began on 10 July 1943. The preparations for the operation were bedevilled by many changes of plan in which the overall ground commander, General Harold Alexander, proved unable to direct his principal subordinates: General Montgomery, commanding British Eighth Army, and General Patton, commanding US Seventh Army. As a result, there was never a clear plan for the capture of the island, and opportunities to win the battle quickly and decisively were missed.

Although many of the accompanying airborne troops landed in the sea and drowned because of poor pilot training and bad weather, the initial Allied landings in the south and south-east of Sicily were a success. The large Italian forces put up little resistance and many surrendered readily – however, the German troops were a different matter. They used the rugged terrain expertly in a series of delaying actions, while the Allied commanders quarrelled over how to conduct the campaign. Finally the Germans withdrew across the Straits of Messina in mid-August virtually unmolested by the superior Allied air and naval forces.

AFTER SICILY

With so many troops and resources committed to the Mediterranean by early 1943, and no possibility of organizing a cross-Channel invasion before 1944, the Allied leaders faced a dilemma. Britain wanted to continue Mediterranean operations and, since the alternative was to have the troops stand idle, the Americans agreed in May 1943 that an invasion of mainland Italy would follow the battles

ITALY, 1943–5

Throughout the campaign in Sicily and Italy the Allied forces had to fight hard for every advance.

for Sicily. Once again there were disputes about planning and priorities that would continue to affect the Italian campaign to the end of the war.

Mussolini had been deposed as head of the Italian government in July and the new regime began secret peace talks with the Allies. On 3 September Eighth Army crossed from Sicily to the toe of Italy and on the 8th the Italian surrender was announced. The Germans were ready, however, and had moved reinforcements into the country to take over. On the 9th the main Allied landings, by General Mark Clark's US Fifth Army, went in around Salerno, just south of Naples, and were nearly thrown back into the sea during the first few days.

SLOW RETREAT

For the rest of the year the Germans fell back slowly from one well-defended river line to the next. Eighth Army pushed up the east side of Italy and Fifth Army to the west. By the turn of the year the Allied advance had reached the Germans' Gustav Line, whose

Below: German paratroops bring supplies to the Cassino defences.

most famous bastion was centred on Monte Cassino, still well to the south of Rome.

In an attempt to break the stalemate the Allied forces made an amphibious landing at Anzio, behind the German lines, on 22 January 1944. The troops there, timidly led, soon found themselves effectively besieged in their beachhead. Repeated attacks on the Gustav Line over the following months also failed.

During May 1944 the Allies at last mounted a properly co-ordinated attack all along the Italian front, and this time they

Above: A US cargo ship explodes off a Sicilian beach following a German air attack, July 1943.

captured Cassino and broke the Gustav Line. By then, however, Montgomery and many veteran troops had left to prepare for D-Day and Italy had slipped down the Allied priority list. Rome fell on 4 June, but by autumn 1944 the Germans were again making a stand, this time on the Gothic Line just north of Florence.

Below: A British Bofors gun in the ruins below Monte Cassino.

Infantry Support Weapons

Although all infantrymen carried personal weapons, success in battle often came from their heavier equipment. Infantry firepower from the fire-team to battalion level depended above all on machine-guns and mortars.

The principal support weapons employed within infantry units were machine-guns and mortars. Both categories included lighter weapons designed to be manoeuvred quickly between locations, and heavier types for use from longer-established positions in both defence and attack.

LIGHT MACHINE-GUNS

Infantry units in all armies worked in squads of roughly ten men with one or more light machine-guns providing the squad's main firepower in both attack and defence. Weapons of this type included the Soviet 7.62mm (0.3in) Degtyarev DP1928, the British 0.303in (7.7mm) Bren and the US 0.3in (7.62mm) Browning Automatic Rifle (BAR), all of them normally

bipod-mounted and magazine-fed. The Bren, highly accurate and reliable, and the Degtyarev, with a very useful 47-round magazine, were both more successful designs than the BAR, which was clumsy in action and had a smaller 20-round

Above: A British Bren gunner firing on German positions at Monte Cassino in Italy in early 1944.

magazine. Minor weapons also included France's 7.5mm (0.295in) Châtellerault.

HEAVY MACHINE-GUNS

Most armies had battalion or regimental support companies equipped with machine-guns for use in the sustained-fire role, aiming to deny areas of ground to the enemy in either attack or defence. These weapons were usually tripod-mounted and physically heavier than the squad light machine-guns, but not necessarily in larger calibres.

In fact the German Army used its MG34 and MG42 both as bipod manoeuvre weapons and on tripod mountings in

Left: A German MG34 on a tripod mount for use as a heavy machine-gun on the Eastern Front in 1941.

3-INCH MORTAR

Britain's 3-inch mortar was used throughout the war by the support companies of British infantry battalions. Versions were produced with varying barrel lengths – 130cm (51in) was standard. A lighter 76cm (30in) barrel, designed for jungle warfare, proved to be rather inaccurate.

CALIBRE: 76.2mm (3in)
WEIGHT: 57.2kg (126lb) – barrel 20kg (44lb), base plate 16.8kg (37lb), bipod 20.4kg (45lb)
RANGE: 2,560m (2,800yd)
BOMB WEIGHT: 4.5kg (10lb)

Various armies also had heavy-calibre machine-guns, often used in a combination of ground and anti-aircraft roles and also commonly fitted to armoured vehicles. Notable types were the Soviet 12.7mm (0.5in) DShK1938 and the American 0.5in Browning M2 HB.

MORTARS

Infantry mortars came in two main calibres: roughly 50mm (2in) and 80mm (3in). Japan's 50mm (1.97in) Type 89 was typical of the smaller weapons, firing a 0.8kg (1.76lb) bomb up to 650m (700yd). Allied troops called it the "Knee Mortar", erroneously thinking it could be fired safely while balanced on a soldier's leg.

Another simple design was Britain's 2-inch mortar, which could fire a 1.1kg (2.4lb) bomb some 500m (550yd). In the larger calibres, several countries used versions of a French Brandt design, which included the American 81mm (3.19in) M1, firing a 4.8kg (10.6lb) bomb 2,250m (2,450yd). The Soviet 82mm (3.23in) PM37 and the Japanese 81mm Type 99 had a similar performance. Most mortars in this class came in three man-portable (but still very heavy) parts: barrel, base plate, bipod.

FLAMETHROWERS

Every major army used man-portable flamethrowers for such specialist tasks as bunker-busting. All had similar capabilities and similar drawbacks. All were very heavy (up to 40kg/90lb), and had a limited range (40–50m/yd) and a modest fuel supply (10 seconds or less).

Operating them was hazardous in the extreme, not least because flamethrower men could expect no mercy if they were captured by an enemy. Probably the most prolific use of flamethrowers was by the US Marine Corps in the battles for the Pacific islands.

Below: A German soldier with a Flammenwerfer 35, a particularly heavy early-war flamethrower design.

the sustained-fire role. The US Army filled this latter requirement with the 0.3in Browning M1917 (also used as a light machine-gun); Britain had the 0.303in Vickers; and the Soviets employed the 7.62mm PM1910 and Goryunov SG43. All were belt-fed and thoroughly reliable, though the PM1910 was particularly heavy.

Other nations' designs were often less satisfactory. The Italian Breda 6.5mm (0.256in) Modello 1930 light machine-gun and the heavier Fiat-Revelli Modello 1935 (in the same calibre) were both very prone to jamming.

Night Bombing of Germany, 1939–44

After the French surrender, aerial attack was the only way in which Britain could strike back directly at Germany. RAF leaders welcomed the development of an all-out bombing campaign; they thought that their service could win the war on its own.

Britain's Royal Air Force (RAF) had been created in 1918 specifically to bomb Germany and held to this aim as war approached in the 1930s. A small number of daylight raids during the Phoney War showed that bombers were hopelessly vulnerable to fighter attack and confirmed that, when attacks began in earnest in 1940, they would be made mainly at night.

In the first phase, up to early 1942, very little was achieved, though losses, compared to some later points in the campaign, were also modest. Bombers were sent to attack specific targets – individual factories or rail junctions, for example – but in fact their navigational skills were so poor that they rarely hit them.

AREA BOMBING

Air Marshal "Bomber" Harris took over at the head of RAF Bomber Command in early 1942 and changed tactics, at that

Above: A German poster urging blackout precautions.

stage with full support from his service and political superiors. Throughout his period in command Harris believed that "area bombing" of the residential districts of enemy cities was the

only sensible use of his force and that, correctly applied, it would win the war.

The aim was to cripple the German war effort by attacking the morale of the working population, killing some and at least dehousing many others. Little thought was given to whether this was a morally justifiable means of waging war.

The first effective attacks in this vein came in the late spring of 1942, most notably a handful of "thousand-bomber" raids that employed all of Bomber Command's training aircraft, as well as the operational force. These raids coincided with the introduction of Gee, the first significant electronic navigation aid, and the beginning of a process whereby the accuracy of attacks was improved. However, Harris foolishly opposed the creation of a Pathfinder Force to lead attacks and mark targets but he was overruled.

Throughout 1942 the aircraft used by Bomber Command also greatly improved, with the four-engined Halifax and Lancaster types with their large bomb loads predominating by early 1943. Thus, the roughly 5,000 bomber sorties flown in June 1942 dropped 6,950 tonnes of bombs. However, only a year later in June 1943, the 5,800 bomber sorties flown dropped well over 15,500 tonnes.

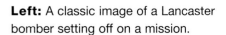
Left: A classic image of a Lancaster bomber setting off on a mission.

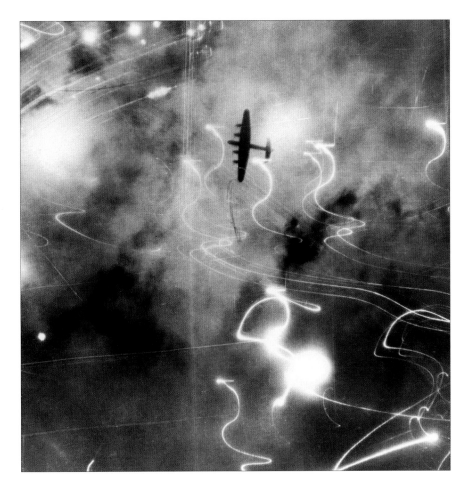

Above: A Lancaster bomber over Hamburg, silhouetted by flares and ground fires in a time-exposure photo taken from another bomber.

In response to these increasingly damaging attacks the Germans built up a formidable night-fighter force and control system. Initially this relied exclusively on ground stations directing twin-engined heavy fighter designs into close enough contact with a target aircraft that their radar systems could complete the interception. Later single-engined day fighters were sent by radio to the skies above the cities being attacked and were able to make visual interceptions.

In response the RAF concentrated its aircraft to swamp particular sectors of the system, as well as jamming radars and using electronic and other deception measures to fool controllers or blind their equipment.

In 1943 Bomber Command devastated many cities across Germany, but the only time it came close to achieving the sort of results Harris expected was in a series of attacks on Hamburg in July. These benefited from a temporary Allied advantage in the electronic warfare struggle and because of this and other special circumstances were never to be repeated.

BATTLE OF BERLIN

In the winter of 1943–4 Harris mounted what he called the Battle of Berlin. However, for all the damage inflicted, this series of attacks on the German capital and other cities was ultimately a defeat for Bomber Command. Losses rose to a rate that could

AIR MARSHAL ARTHUR HARRIS

Harris (1892–1984) led RAF Bomber Command from February 1942 to the end of the war. Harris fervently believed that, suitably strengthened, his command could win the war on its own by bombing Germany's cities. His leadership undoubtedly transformed the morale and effectiveness of Bomber Command, which were at a very low ebb when he took over. However, his forces lost increasingly heavily in 1943–4 and only their partial diversion to support operations in France saved them from an even greater defeat.

Above: Air Marshal Harris (seated) and senior members of his staff at Bomber Command.

not have been sustained for long, while the German economy and civilian morale never came close to collapse. Though Harris complained bitterly and disobeyed his orders as much as possible, it was a respite for Bomber Command to be switched partly to targets in France over the next months, in preparation for D-Day.

Night Fighters

During WWII radar developments gradually stripped away the cover of darkness from night air operations. Starting with the Blitz in 1940–1, night fighters took an increasing toll of attacking and defending aircraft in a bitter struggle for supremacy.

No night-fighter aircraft existed anywhere at the beginning of WWII; night air defence relied on an inadequate combination of searchlights and anti-aircraft guns. However, by the later stages of the 1940–1 Blitz, Britain had introduced the first effective night fighters, directed by ground radar to a position near the target aircraft and then closing in to attack, using their own airborne radar equipment. Most night fighting would follow this pattern for the remainder of the war, though the radars used grew in range and precision, and countermeasures to defeat them became more sophisticated.

BRITISH DESIGNS

The Bristol Beaufighter was the first successful night fighter; like most night fighters of the war, it was a two-seat, twin-engined design. Twin engines left the aircraft nose free for the radar equipment (and usually heavy guns as well) and gave the power necessary to overcome the drag often caused by bulky aerials; the second crewman was the radar operator. The Beaufighter made its operational debut in September 1940 and achieved its first success over the next month.

From 1941 it was joined and eventually replaced by a series of variants of the De Havilland Mosquito. This was fast and long-ranged and very heavily armed – commonly four 20mm (0.79in) cannon and four machine-guns – and equipped with successively improved models of radar. Late-war versions also carried Serrate equipment to home in on German night-fighter radar signals. By then night-intruder operations over enemy territory were the main mission for the Allied night-fighter force.

Since there was little likelihood of night air attack on the USA and the US bomber forces operated by day, American night-fighter development was rather slower. Initially US night-fighter units used the Beaufighter and versions of the A-20 Havoc bomber, but they were converted from mid-1944 to the Northrop P-61 Black

NORTHROP P-61

The P-61 Black Widow first flew in 1942 and was first deployed on operations in June 1944, in both the Pacific and Europe. Some 740 were built in all. Some variants had only the nose-mounted cannon and two crew; others had a fully trainable turret with two or four machine-guns.

CREW: 2 or 3
ENGINES: 2 x Pratt & Whitney R-2800-65; 2,250hp each
SPEED: 589kph (366mph)
CEILING: 10,100m (33,100ft)
ARMAMENT: 4 x 20mm (0.79in) cannon + 4 x 0.5in (12.7mm) machine-guns

HEINKEL 219

Germany's chaotic air procurement system is illustrated by the He 219, only produced after significant private investment by the manufacturer. Speed and altitude performance were generally much better than earlier German types, but it compared poorly with the Mosquito or P-61.

CREW: 2
ENGINES: 2 x Daimler Benz 603
SPEED (A-7): 616kph (383mph)
RANGE: 1,540km (960 miles)
CEILING: 9,300m (30,500ft)
ARMAMENT: up to 4 x 3cm (1.18in) + 4 x 2cm (0.79in) cannon

Above: A German Ju 88 night fighter fitted with the clumsy aerial array of the Lichtenstein SN-2 radar.

Widow. This was the only purpose-built night fighter to see service with any nation during the war and had virtually identical performance characteristics and armament to the Mosquito.

The US Navy also used a number of Corsair and Hellcat single-seat, single-engined fighters at night, fitted with radar sets in the wings.

GERMANY'S REPLY

Since it faced the most sustained night-bombing attacks of the war, Germany naturally responded with significant night-fighter developments, though these were hampered by the poor organization of German aircraft procurement and electronics research.

Early types included versions of the Messerschmitt (Me) Bf 110 fighter and Junkers

(Ju) 88 bomber. Both could carry a heavy armament, including the *Schräge Musik* ("Jazz Music") upward-firing cannon used from the blind spot underneath a target aircraft. However, performance suffered because of the drag from the large aerials required by the German Lichtenstein radar sets.

A variety of other models were also used, including the Me 210 and 410, and the Ju 188 and 388. The most capable of all, but produced in limited numbers (fewer than 300), was the Heinkel 219, in service from the summer of 1943.

Germany also made extensive use of unmodified single-engined day fighters in the night-fighter role. Since the British night bombers operated in concentrated streams, a fighter directed to the stream had a reasonable chance of acquiring a target visually, especially close to the bombing target when flares and ground fires gave

BRISTOL BEAUFIGHTER

Developed from the Beaufort torpedo bomber, the Beaufighter first flew in July 1939. It was used as a radar-equipped night fighter from the autumn of 1940 but was gradually replaced in this role by the Mosquito. Torpedo-bomber and strike variants saw extensive use to the end of the war.

CREW: 2
ENGINES: 2 x Bristol Hercules III; 1,400hp each
SPEED: 540kph (335mph)
RANGE: 2,400km (1,500 miles)
CEILING: 8,800m (28,900ft)
ARMAMENT: 4 x 20mm (0.79in) cannon

extra illumination. This *Wilde Sau* ("Wild Boar") tactic was introduced initially when British counter-measures blinded the German radar control system for a time in mid-1943. This was used with success for the remainder of the war.

Japan had a limited night-fighter force, in part because of the lack of effective Japanese radar equipment. The Navy's Nakajima J1N "Irving" had a small number of successes over the Home Islands and elsewhere, and the Army's Kawasaki Ki-102 "Randy" was potentially a capable aircraft but only appeared in modest numbers.

US Strategic Bombing, 1942–3

Formations of American Flying Fortress and Liberator bombers began ranging over Europe in mid-1942. Their leaders hoped to bring the Luftwaffe and the German economy to their knees, but their initial efforts were bloodily defeated.

Like Britain's RAF, the United States Army Air Force (USAAF) entered the war committed to a policy of long-range strategic bombing. Like the RAF in 1939, it believed in making daylight attacks on precise targets, relying on the heavy defensive armament of its bomber aircraft and their tight formations to ward off enemy fighters. Like the RAF it would also discover that bombing accuracy never approached the anticipated results and that bomber formations could not defend themselves adequately against fighter attack.

The principal USAAF heavy-bomber organization throughout the campaign was the Eighth Air Force (AF), based in Britain. The Ninth and Twelfth AF, based in North Africa in 1942–3, also included heavy-bomber units, which were transferred to

Above: B-17 *Memphis Belle* and its crew on their return Stateside after completing their 25-mission tour of duty, one of the first crews to do so.

the Fifteenth AF, operational in southern Italy from the end of October 1943.

The first raid over western Europe by a wholly American bomber force was on 17 August

1942 against targets at Rouen in northern France. For the remainder of that year all targets attacked were similarly short-range and the bomber forces were provided with strong fighter escorts, both from the RAF and the American forces. The build-up of the American bomber strength was slow. By the end of the year only some 1,550 sorties had been flown, with 32 aircraft lost.

ATTACKING GERMANY

The first raid on a German target was against Wilhelms-haven on 27 January 1943; 91 aircraft attacked, with 3 shot down in return for 6 German fighters. This would prove to be an unusually favourable ratio of losses for the Americans.

For the next several months other German ports were also targeted along with French ports used as U-boat bases. The raids on the French ports, like those conducted by the RAF in the same period, were a fiasco. They flattened the homes of the local French population but barely scratched the thick concrete of the U-boat pens.

Eighth AF first flew over 2,000 sorties in June 1943 and would pass the 6,000 mark monthly in January 1944. The attacks by the British-based

Left: A B-24 Liberator on fire after being hit by flak. Watching comrades die like this was difficult, to say the least, for bomber crews.

Above: A formation of B-17s on a raid over Germany. Bombing accuracy in cloudy European skies never matched expectations.

Above: Bombs dropping away from a B-17 during a raid on Bremen. Other attacking aircraft leave condensation trails below.

forces were joined, from the early summer of 1943, by raids on targets in Italy and elsewhere from North Africa.

Although the attacks up to mid-1943 had not yet involved concerted attempts to penetrate deep into German airspace, they had brought about a significant strengthening of the German home-defence force, from about 450 to 1,100 day fighters. Despite Anglo-American efforts to bomb aircraft factories, German fighter production increased through 1943 right up to September 1944.

HEAVY LOSSES

The first big tests of the US commanders' belief that their bombers could fight their way to their targets unescorted came in August 1943. On the 1st the Mediterranean force lost very heavily in a raid on the oilfields at Ploesti in Romania. On 17 August some 376 US bombers attacked the Messerschmitt factory at Regensburg and the ball-bearing manufacturing plant at Schweinfurt; 60 bombers were lost and Schweinfurt in particular was little damaged.

Even though part of the bomber escort was made up of P-47 Thunderbolts fitted with longer-range fuel tanks, these could reach only just inside Germany; most of the losses were on the long unescorted leg of the mission between there and the target and back again.

Confirmation of the desperate need for longer-range escort fighters came in October. A series of raids, including one on Schweinfurt on the 14th, cost

Eighth AF about half its operational strength. The American commanders abandoned deep-penetration attacks for the moment. The Luftwaffe had won the first round of the battle.

Right: General Ira Eaker, seen in 1942. Eaker commanded Eighth Air Force during its unsuccessful operations in 1943.

Anti-aircraft Guns

Although a single rifle bullet could bring down an aircraft, effective anti-aircraft fire usually depended on a combination of automatic weapons and slower-firing medium guns with greater destructive power and range.

In the course of WWII many thousands of aircraft on all sides were brought down by fire from the ground. High-flying heavy bombers or long-range fighters and low-altitude ground-attack aircraft were all vulnerable, though usually to different weapons. As in most other classes of land-warfare weapons, the types of anti-aircraft (AA) guns in service changed little during 1939–45, though ammunition and control equipment developed substantially.

LIGHT AA GUNS

Since lower-flying aircraft appear to be travelling faster to an observer on the ground, weapons to shoot down such planes have to be capable of traversing quickly and firing multiple shots rapidly, since a target might only be in sight for a few moments. In practice this meant weapons of roughly 40mm (1.5in) or less, typically firing a shell weighing less than 1kg (2.2lb) to an effective ceiling of up to 3,500m (11,500ft).

Although soldiers could and did attempt to engage aircraft with firearms of every kind up to and including standard machine-guns on specially adapted mounts, the smallest purpose-built AA system in widespread use with a major army was the American "Quad Fifty". This was a quadruple mounting carrying four 0.5in (12.7mm) M2 Browning heavy machine-guns, which appeared both on a towed trailer and on various self-propelled mounts. However, this was never entirely satisfactory as the individual rounds lacked sufficient striking power to bring down an enemy aircraft reliably.

The Germans had a 2cm (0.79in) weapon (and also developed a four-gun mount) but this had similar shortcomings. Japan, Italy and Britain had similar single-barrel 20mm weapons. Some of the British and Japanese weapons were based on a design originating with the Swiss Oerlikon company.

The next step up for Germany and the USA was to 3.7cm (1.46in), a calibre also used by the Soviets. Britain's main light AA gun was the 40mm (1.58in), built under licence from the Swedish Bofors company, a weapon also used by many other combatant nations. This fired a 0.9kg (2lb) high-explosive shell at a practical rate of 80–100rds/min.

Like most guns in this class the Bofors was usually fitted with simple visual sights, which were often the only ones used. Like other guns it also had various

---

M3 3-INCH GUN

The M3 3-inch gun was the standard US medium anti-aircraft gun in the early part of the war but was gradually replaced by the M3 90mm. Both guns were also used in ground roles, like this 3-inch shown in New Guinea in 1943.

CALIBRE: 3in (76.2mm)
SHELL WEIGHT: 5.6kg (12.3lb)
RATE OF FIRE: 25rds/min
EFFECTIVE CEILING: 9,100m (30,000ft)
MUZZLE VELOCITY: 853m/sec (2,800ft/sec) – HE shell

Right: A German Flakvierling quad 2cm (0.79in) Flak 38, in northern France in 1944.

The 40mm Bofors gun was designed in the early 1930s and was used by many combatant nations in WWII, both on land and as a ship-borne weapon. A British Army example is illustrated but it was also produced in twin- and quad-barrel versions for naval use.

Above: A British 3.7in (94mm) gun in action at night. Like many other British AA batteries, this one had a mixed male and female complement.

CALIBRE: 40mm (1.58in)
SHELL WEIGHT: 0.9kg (2lb)
RATE OF FIRE: 160rds/min max.
ABSOLUTE CEILING: 6,800m (22,300ft)
MUZZLE VELOCITY: 880m/sec (2,890ft/sec)

mechanical predictor sights (the type depending on the country), designed to help the gunners allow sufficient "lead" ahead of a fast-moving target.

HEAVY AA GUNS

Small numbers of guns of around 127mm (5in) were used by various nations, including the German 12.8cm (5.04in) FlaK 40, the US 120mm (4.72in) M1 and Britain's 5.25in (135mm). However, these were at the point where the gain in ceiling and striking power from the larger calibre began to be outweighed by slow rate of fire and clumsiness in action.

More common were lighter weapons similar to Germany's "Eighty-eight", various versions of an 8.8cm (3.46in) gun firing a 9.4kg (20.7lb) shell to over 8,000m (26,200ft). Britain's 3.7in (94mm) and the US 90mm (3.54in) weapons were broadly comparable. Like the Eighty-eight the 90mm on the M2 mount could be used as an anti-tank weapon.

HITTING THE TARGET

In the early-war years all AA guns had to rely, at best, on mechanical predictors calculating where to aim by using human estimates of aircraft height, speed and course.

During the war these were gradually replaced by radar systems which, among other advantages, could be used suc-cessfully at night or through cloud. Heavier AA shells initially relied on time or barometric fuses which, when set using radar information, could indeed be extremely accurate.

Better still was an Anglo-American development, the proximity fuse, in effect a radar set that could be fitted in a shell to detonate it when it went close to a target. This was used with great success later in the war against Japanese kamikazes and German V-1 missiles.

Heavy Bombers

The four-engined heavy bombers of the British and US air forces were among the most potent weapons of the war. They combined high-tech electronic equipment with the brute power to deliver tonnes of bombs deep inside an enemy country.

From the later years of WWI airforce officers in various countries had argued that, if suitably equipped and expanded, their services could win a war by bombing attacks on the industries and people of an enemy homeland. Such attacks were called strategic bombing (as distinct from tactical operations near a land battle front).

In WWII Britain and the USA were the only countries to equip themselves for such operations and try to win the war by carrying them out. At the heart of strategic bombing campaigns were the various four-engined bombers described here.

The first aircraft of this type to enter service was the Boeing B-17 Flying Fortress, first flown

Above: A pair of B-24 Liberators escorted by P-40 fighters on a mission in the north Pacific in 1944.

in 1935 and in series production from 1939. This had impressive performance but, with only five hand-trained machine-guns and lacking self-sealing fuel tanks and adequate armour protection for the crew, early marks hardly lived up to their name. Significant improvements came

with the B-17E version, in use from 1942; the final B-17G carried 13 defensive guns.

At first sight the B-17's near contemporary, the Consolidated B-24 Liberator (first flown in 1939), was superior. It was faster, and had a better range, with the same bomb load and similar defensive armament. However, in action it was found to be less suited to the tight formation flying at high altitude needed for operations over Germany. Like the B-17 it went through various marks before reaching its best defended final version, the B-24J.

DAY BOMBING

Throughout the war in Europe US tactics were to carry out strategic bombing raids by day in the hope of bombing with great accuracy. This dictated the heavy defensive armament fitted to the B-17 and B-24 and

---

BOEING B-17G FLYING FORTRESS

The B-17G was the definitive version of this aircraft. About two-thirds of the 12,700 B-17s were of this model. Improvements on the original included a lengthened fuselage, larger tail, and chin and tail turrets.

SPEED: 462kph (287mph)
ENGINES: 4 x Wright Cyclone R1820 radials; 1,500hp each
RANGE: 3,200km (2,000 miles)
ARMAMENT: 13 x 0.5in (12.7mm) machine-guns; typically 2,000kg (4,400lb) bombs

PIAGGIO P.108B

Italy's P.108B was one of the few heavy bombers outside the UK and US forces to see action. The P.108B was a powerful and effective machine with various advanced features, including remotely controlled gun turrets.

Speed: 430kph (267mph)
Engines: 4 x Piaggio PXII RC35 radials; 1,500hp each
Range: 3,540km (2,200 miles)
Armament: 6 x 12.7mm (0.5in) + 2 x 7.7mm (0.303in) machine-guns; 3,500kg (7,700lb) bombs

also resulted in their having relatively modest bomb loads of some 2.5 tonnes. In the event bombing accuracy in cloudy European conditions was much less than in Stateside trials and, until the advent of long-range escort fighters in 1944, bomber losses were extremely heavy.

BRITISH TACTICS

Britain's RAF began the war equally convinced that its bombers could fight their way to their targets in daylight. However, bitter experience soon proved otherwise and Bomber Command switched to night raids on Germany.

Though twin-engined types featured in these attacks into 1943, four-engined designs to replace these entered service from early 1941. Relying on the cover of darkness, all had weaker defensive armament than the US types but much greater bomb loads. Bombing accuracy was very poor in the early stages of the campaign, but it improved substantially later as electronic aids to navigation were introduced and improved.

The first type, the Short Stirling, only saw front-line use until 1943 – its low service ceiling made it unacceptably vulnerable. The Handley Page Halifax, which came next, was a significant improvement, especially the later Mark 3 type, and served to the end of the war. Undoubtedly the best British heavy bomber of the war, how- ever, was the Avro Lancaster, in squadron service from early 1942. Over 7,000 of this tough and reliable aircraft were built.

Taking the war to Japan demanded a longer-range air- craft than those deployed in Europe, and design work on what became the Boeing B-29 Superfortress began before the war. The prototype first flew in 1942 but, between then and its combat debut in mid-1944, there were many problems and modifications to be addressed. This was largely because, in addition to its great size, it was an extremely complex and tech- nologically advanced aircraft. However, with the develop- ment of bases in the Mariana Islands later in 1944, it was able to begin attacks on Japan, cul- minating in the nuclear missions that finally ended the war.

Below: A flight of Short Stirling bombers during training in 1942. The Stirling was the least effective of the three British four-engined designs.

Clearing the Ukraine

*Between mid-summer 1943 and late spring 1944 a series of crushing Soviet attacks
drove the Germans out of the Ukraine, pushing them back into pre-war Poland
and Romania. In the north the 900-day siege of Leningrad was also lifted.*

From the conclusion of the Kursk offensive to the end of the war a little less than two years later, the Red Army would advance relentlessly and almost continuously for victory.

In the first phase of the battles the Soviets pushed forward all along the southern two-thirds of the front, to reach and cross the Dniepr River and recover such major cities as Smolensk, Kiev and Dnepropetrovsk. In the second phase, in the winter and early spring of 1944, there

Above: German soldiers in the cold and snow of an Eastern Front winter, March 1944.

were attacks in the north and south. In the north the bitter siege of Leningrad was finally broken; in the south the rest of the Ukraine was recaptured and the first steps were taken toward the conquest of southeastern Europe.

TO THE DNIEPR

The process began with Soviet attacks on the north of the Kursk salient on 12 July and was extended by a further offensive on the south flank of the salient on 3 August. Although the Germans defended very skilfully against these advances, they had to give up Orel to the northern drive in early August and Kharkov to the southern attack on the 23rd.

By the end of August Soviet forces were advancing everywhere south of Moscow – but especially successfully in a broad area west of Kursk and in another sector in the far south toward Stalino (Donetsk).

Aware of the possible Italian surrender (which became a reality on 8 September), Hitler had already begun to reinforce his troops in Italy. Allied landings on mainland Italy in early September made this still more urgent. He therefore gave permission for the German troops in the eastern Ukraine to retreat

SOVIET VICTORIES, 1943–4
The Red Army's advances cleared the Germans from most of the territory of the pre-war USSR.

Map

Gulf of Riga

• Riga

■ MOSCOW

USSR

• Smolensk

• Minsk

• Orel

• Bialystok

• Kursk

Mozyr

Brest •   Sarny

• Belgorod

Kiev •   Kharkov •

• Lubny

Lvov •   Vinnitsa

• Izyum

Voroshilovgrad •

• Rostov

Sea of Azov

Odessa • Perekop •

Novorossiysk •

Sevastopol • • Yalta

Black Sea

0   100   200   300 mi
0   200   400 km

Soviet attacks
German attacks
Front line 4 July 1943
Front line 30 Sept 1943
Front line 23 Dec 1943
Front line 24 Jan 1944
Front line mid-Apr 1944

to the Dniepr. As they did so they converged on the various major towns with bridges over the river. The Soviets advanced into the gaps between these and had improvised crossings themselves by the end of the month north of Kiev and west of Dnepropetrovsk. Kiev itself was taken in early November.

ON TO THE BORDERS
By December, despite the able defensive leadership of Field Marshals Erich von Manstein and Ewald von Kleist, the German front included a number of vulnerable salients, held on Hitler's orders. In attacks

Above: Red Army soldiers receive a warm welcome from local people after the recapture of Kiev.

beginning on 24 December and extending through March 1944, these were eliminated at the start of new advances. Both field marshals joined the growing list of able German generals dismissed for making withdrawals and arguing too often with their erratic *Führer*.

When this phase of fighting ended in mid-April, the front line ran from the Pripet Marshes via the eastern foothills of the Carpathian Mountains to the Black Sea coast west of Odessa.

SIEGE OF LENINGRAD

In terms of its effect on the outcome of the war, the siege of Leningrad (St Petersburg) was a sideshow at most. However, with probably more than a million civilians dead, most by starvation, disease and cold, it was a human tragedy on a vast scale.

By early September 1941 the Germans' first advances had cut Leningrad off from the rest of the USSR, and Hitler decided to obliterate the city by bombing, shelling and starvation. The Soviet authorities had delayed evacuating the civilian population of 2.5 million: about a third were evacuated in 1942; a third remained in the city, fighting or working in the factories; and a third died in the first dreadful winter.

For more than a year, despite attacks and counter-attacks by both sides, a trickle of supplies (but only that) reached the city across Lake Ladoga. The supplies travelled by boat in summer and by truck over the ice in winter, but all the time under German shelling and air attack. New Soviet attacks in January 1943 opened up a narrow land route into the city, easing the worst privations, but got no further. Finally, in January 1944, a new offensive forced a general German retreat. Stalin declared the city liberated on the 27th.

Left: German troops on the retreat with a horse-drawn 5cm (1.97in) PaK 38 anti-tank gun in the Ukraine in the mud of the spring thaw, 1944.

Heavy Artillery

The big guns of the heavy artillery were among the most fearsome land warfare weapons of WWII. They also appeared by the hundred, or more – the Soviets used over 16,000 guns of all calibres in their final attack on Berlin in 1945.

Josef Stalin is said to have described artillery as the "god of war". No soldier on the receiving end of a bombardment from the heavy guns of any major army would have been likely to disagree. Heavy artillery weapons were usually allocated to higher formations (corps and armies or similar) and would be capable of switching their support from unit to unit, both in defence and in attack.

ARTILLERY TACTICS

As in other categories of artillery weapon, the Soviets were the most prolific users of heavy guns. Britain and the USA had the most sophisticated organization, able to shoot elaborate suppressive bombardment plans and to respond rapidly to events with stunning concentrations of firepower.

There were numerous weapons of this class in use: Germany had over 200 types of artillery weapon (of all calibres) in service. To give a more particular example, the Soviets had at least five models of 152mm (5.98in) howitzer and two 152mm guns. Accordingly only a representative sample can be discussed here.

Most armies had weapons closely comparable to the Soviet designs just mentioned, firing a shell of roughly 45kg (100lb) to a range of 15km (9.3 miles) for higher-trajectory howitzers or up to 27km (17 miles) for flatter-trajectory guns. Along with their different firing characteristics, the howitzers were also lighter overall and, therefore, usually more mobile and normally cheaper and easier to build, by no means a trivial consideration.

Britain's 5.5in (140mm) gun was typical. It fired a standard 45.4kg (100lb) shell to 14.8km (9.2 miles), or a 37kg (82lb) shell to 16.6km (10.3 miles). A little over 6 tonnes in action, it had a maximum rate of fire of perhaps 3 rounds a minute. The USA's 155mm (6.1in) Gun M1 fired a slightly lighter shell to a range of over 23km (14.3 miles), but the gun and mounting were twice as heavy overall as the 5.5in. Lesser-known but also effective types included Italy's 149mm (5.87in) Cannone da 149/40 M35 and France's 155mm M1932 Schneider.

SUPER-HEAVY WEAPONS

Most nations also included heavier guns and howitzers in their armoury. The US 8in (203mm) Gun M1 came into service mid-war. It could fire a 109kg (240lb) shell 35.5km (22 miles) on long-range-bombardment and counter-battery missions. Bigger still was its contemporary the 240mm (9.45in) Howitzer M1 with a 157kg (346lb) shell and a range of 23km (14.3 miles).

Left: A US 155mm (6.1in) M1 "Long Tom" gun, in action on Leyte during November 1944. The split-trail design gave a very stable firing platform.

5.5-INCH GUN

The British 5.5in (140mm) gun was introduced in 1940 and saw widespread service (the example shown is in Italy in 1943). There was also a slightly longer-ranged 4.5in (114mm) weapon mounted on the same carriage.

CALIBRE: 5.5in (140mm)
WEIGHT IN ACTION: 6.3 tonnes
SHELL WEIGHT: 45.4kg (100lb) or 37kg (82lb)
RANGE: 14.8km (16,200yd) – 100lb shell; 16.5km (18,100yd) – 82lb shell
MUZZLE VELOCITY: 619m/sec (2,030ft/sec) max.
CREW: 10
RATE OF FIRE: 3rds/min max.

Railway guns were a particular German speciality including the longest-range and heaviest-calibre weapons to see action during the war.

The best-known of these was the 28cm (11in) K5 (E), employed against the Allied beachhead at Anzio among other places. This fired a 255kg (562lb) shell 63km (39 miles) or up to 86.5km (54 miles) with rocket assistance. A 28cm gun was reworked to 31cm (12.2in) calibre for test-firing a fine-stabilized round an even more astonishing 150km (93 miles).

The biggest weapon of all was a massive 80cm (31.5in) calibre. Two such guns were built, but only one is known to have been used (in the German siege of Sevastopol in 1942). The gun weighed some 1,350 tonnes in action, moved on two sets of railway tracks and fired a 7-tonne high-explosive shell up to

47km (29 miles). The design and industrial effort required to produce such a weapon was wholly disproportionate to its highly limited combat worth.

Below: A battery of German 15cm (5.91in) Kanone (E) railway guns in position near the French border in 1940. In fact few of these weapons saw action.

The Soviet Invasion of Germany Fighting in the German town of Küstrin in early 1945.

Totalitarian Rule, Germany and the USSR Inmates of the Sachsenhausen concentration camp.

Resistance Female fighters of the French resistance greet the liberation of Marseille in 1944.

Under German control 7 May 1945
Neutral
Front line 23 June 1944
Front line 26 Aug 1944
Front line 15 Dec 1944
Front line 22 Mar 1945
Front line 7 May 1945

0   100 200 300 400 500 mi
0   200   400   600   800 km

ICELAND

ATLANTIC OCEAN

NORWAY

SWEDEN

FINLAND

North Sea

DENMARK

Baltic Sea

ESTONIA

LATVIA

LITHUANIA

EAST PRUSSIA

USSR

IRELAND

UNITED KINGDOM

NETHER-LANDS

BELGIUM

GERMANY

POLAND

CZECHOSLOVAKIA

FRANCE

SWITZER-LAND

AUSTRIA

HUNGARY

ROMANIA

Caspian Sea

PORTUGAL

SPAIN

Corsica

ITALY

Sardinia

YUGOSLAVIA

ALBANIA

BULGARIA

Black Sea

GREECE

TURKEY
Declared war on Germany February 1945

Mediterranean Sea

Sicily

CYPRUS (British)

SYRIA

IRAN

IRAQ (British)

TUNISIA (French)

ALGERIA (French)

LIBYA (Italian)

EGYPT (British)

PALESTINE (British)

TRANS-JORDAN (British)

VICTORY OVER GERMANY

By early 1944 Hitler's Kriegsmarine had been reduced virtually to impotence in the Battle of the Atlantic and that spring the Luftwaffe was decisively defeated in a titanic series of air battles over Germany. Yet, at the start of June, German soldiers stood everywhere on foreign soil and even still maintained a toehold inside the pre-war USSR. By the early autumn the Western Allies had liberated France, Hitler's armies in the east had suffered the devastating defeat known as the destruction of Army Group Centre, and among his erstwhile allies Romania had changed sides and Finland made peace.

Hitler himself, weakened and shaken by his injuries in the Bomb Plot assassination attempt, was increasingly out of touch with reality, while the German economy and its industries were heading into their final collapse. But, despite all that, the German Army remained a formidable and resourceful organization, capable of desperate defence and last-ditch counter-attacks. The true nature of the Nazi regime was finally made plain to all with the liberation of the concentration camps in the spring of 1945, but the European war would not end until almost all of Germany was in Allied hands and Hitler was dead in the ruins of Berlin, capital city of his vaunted "thousand-year *Reich*".

Fighters, 1943–5 A Supermarine Spitfire Mark 14 fighter in flight, the final major production version.

The Battle for Normandy US Army artillerymen in action on the Cotentin Peninsula.

The Fall of Berlin Soviet troops raise their flag over the Reichstag building at the centre of the city.

D-Day

Operation Overlord had been the centrepiece of Anglo-American plans since 1942 and was the largest combined land–sea–air operation in history. The Germans ultimately would have no answer to the Allied firepower and strength.

By the end of 6 June 1944 the Allies had landed some 150,000 men in Normandy. All five landing beaches were secure and everywhere the troops were pushing inland, though not as far as had been hoped and planned. The first day of Operation Overlord had been at least a qualified success.

THE PLANS

Although the Allies had a crushing superiority in the air and at sea, the actual landing forces were not overwhelmingly strong because of limited numbers of landing craft and paratroop aircraft. Several strategies were used to offset this. Months of intensive air attacks had been

Right: German troops building beach defences before D-Day run for cover from an Allied aircraft.

KEY FACTS

PLACE: Lower Normandy, France

DATE: 6 June 1944

OUTCOME: Allied air and seaborne landings succeeded despite strong German resistance.

---

carried out and would continue against roads, rail lines and bridges all across France, with the aim of making it difficult for the Germans to move reinforcements to Normandy.

At the same time an elaborate deception plan had been set up to convince the Germans that the Allies planned to land farther east in the Pas de Calais region. This continued after the landings to suggest that Normandy was a feint and the real Pas de Calais operation was yet to come. Both phases were extremely successful.

The Allied Supreme Commander, General Dwight D. Eisenhower, led over 3 million men with 13,000 aircraft, 2,500 landing craft, 1,200 warships and a range of new equipment. This included the obstacle-crossing and bunker-busting tanks of the British 79th Armoured Division, used only on the Anglo-Canadian beaches, as well as amphibious tanks used by all attack formations. Follow-up forces would benefit from the Mulberry Harbours and have their fuel needs supplied by PLUTO (Pipe Line Under The Ocean).

Although the Germans were well aware that there would be an invasion, they had no

D-DAY PLANS

The Allied attack forces came from bases across the British Isles. German forces were inevitably spread out along the French coast.

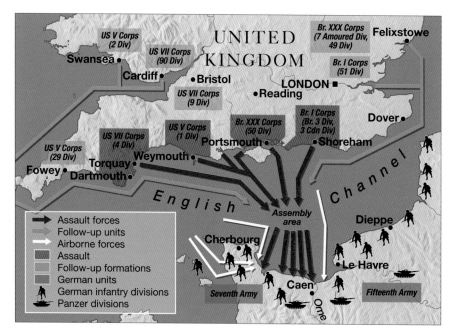

Above: A British commando unit on Sword Beach waits for the order to advance inland.

Above: US Army troops landing on the bitterly contested Omaha Beach on the morning of D-Day.

accurate intelligence of its location or strength. Field Marshal Erwin Rommel now commanded the German forces in northern France. He believed that his only chance was to defeat the invasion before it got properly ashore. He wanted to spread his reserves along the coast so that they could immediately attack any landing forces, because he knew the Allied air forces would make it very difficult to redeploy more distant units.

His superior, Commander-in-Chief West Field Marshal Gerd von Rundstedt, wanted a strong central reserve that would only be sent into the attack once it was clear where the main Allied landings were taking place. The Germans ended up with a compromise: some reserves were near the Normandy coast, but they were not allowed to be deployed without permission from Hitler.

THE LANDINGS

There were five landing areas – Utah, Omaha, Gold, Juno and Sword, from west to east. The first two involved units of US First Army and the last three

British and Canadians of British Second Army. Two US airborne divisions landed by parachute and glider inland from Utah and one British airborne division on the east flank of Sword.

The airborne divisions took most of their objectives and succeeded in disrupting possible enemy counter-attacks. German resistance was fiercest, and Allied casualties heaviest, on Omaha Beach. For a time it looked as if the landing there might fail, but by the end of the day the beach had been cleared. Utah was the easiest of all – with the other three beaches somewhere in between.

Above: One of the vast armament dumps in the UK created by the Allies in preparation for D-Day.

It was important for the Allied troops to grab as much territory as possible at the outset, even if only to make room for the vast follow-up forces. However, the inland advance later in the day came up well short of the D-Day objectives of Caen and Bayeux. Much hard fighting to expand the beachheads clearly lay ahead.

MULBERRY HARBOURS

Since the speed with which supplies and reinforcements could be landed was crucial, and capturing a port would be difficult, the Allied planners decided to solve the problem by taking their own "Mulberry Harbours" with them. Huge breakwaters, causeways and piers were built in Britain, floated across the Channel and then sunk in place off Normandy in the first couple of weeks after D-Day. Two Mulberries were built. One was wrecked in a storm on 19 June, but the other remained in use until late 1944.

Special Purpose Armoured Vehicles

Clearing obstacles and breaching enemy defences were traditionally the tasks of army engineers. Modified tanks and other vehicles designed to carry out these roles were an important factor on D-Day and in other battles.

Battles to overcome elaborate enemy defences as part of an ongoing land campaign, and during waterborne landings at the start of one, were a recurring feature of the Allied counter-offensives in the second half of WWII. Clearing mines, crossing ditches and rivers, as well as destroying enemy strong points were among the tasks required. In line with the mechanization of many aspects of warfare, specialized armoured vehicles to undertake these duties saw significant service – most famously the "Funnies" of the British 79th Armoured Division, which saw action during the D-Day landings and thereafter.

MINE CLEARING

Two main mine-clearing tank types were developed: flails and rollers. Flails carried a rotating drum fitted with chains on the front that beat the ground as the tank drove forward, exploding

Below: DD Sherman tanks during a training exercise. The second tank has not yet lowered its flotation screen, but others have and are immediately ready to fight.

Above: A German Bergepanzer 3 armoured recovery vehicle assists a broken-down Panther battle tank.

mines in its path. These were first used in Matilda Scorpion form at the Battle of El Alamein in October 1942 and, later in the war, appeared in improved Sherman Crab versions. The US Army developed T1 mine rollers that featured an arrangement of steel discs pushed ahead of the tank.

OBSTACLE CROSSING

Ditches and walls were also often encountered. Some tanks were equipped to carry fascines

or similar material, which could be dropped into the gap or provide a ramp. Other tanks, notably the Churchill ARK, simply drove into the ditches or up to the walls themselves and then extended ramps in front and behind, so that subsequent vehicles could just drive over the top. The Carpet or Bobbin tanks unrolled a drum of matting over sandy beaches or similar difficult ground – and this provided a roadway for following wheeled vehicles.

All the major armies of the European war also had various forms of bridge-laying equipment. Germany deployed a

Above: A US T1 Mine Exploder in France in 1944. The 18-tonne rollers could be pushed in front of the Sherman tank at 8kph (5mph).

Above: A Churchill AVRE in late 1944, showing its spigot mortar and obstacle-crossing fascine.

small number of Bruckenleger 4 vehicles, derived from the Panzer 4, in France in 1940, but they were little used. Britain and the USA had Churchill and Sherman bridge-layers.

Perhaps the most important vehicle in this class was the British Churchill Armoured Vehicle Royal Engineers (AVRE). This could carry and position fascines or a short bridge and in addition mounted a spigot mortar to fire demolition charges at pillboxes or other enemy positions. Over 500 AVREs were built; they were extensively used in 1944–5.

NORMANDY BATTLES
One adaptation relatively seldom used, other than on D-Day, was also significant. These were the so-called Duplex Drive (DD) tanks – ordinary battle tanks that were made amphibious by being fitted with canvas screens to raise around the hull

Right: A Churchill Carpet during training for D-Day. The Carpet's matting could help wheeled vehicles cross barbed-wire obstacles.

and a propeller to drive them through the water at some 7kph (4mph). As soon as they reached the landing beach, the tanks would simply drive out of the water, drop the screens and then they would be able to fight entirely normally. The DD concept was originally tested on the Valentine tank from 1941, but by 1944 Shermans were invariably used.

Whether the DD tanks made much difference on D-Day is a debatable issue and relatively few of these were employed. However, a few weeks later a simple improvised modification applied to most of the tanks of US First Army may have made a real difference. This was the Rhino, a set of tusk-like prongs welded to the front of the tank to enable it to cut through hedgerows, rather than being confined by the many high banks that enclosed Normandy's roads.

The Battle for Normandy

It took two months of bitter combat before the Allied forces were able to break out of their Normandy landing areas. The Allies had overwhelming resources and dominated the air, but German battlefield skill made them fight all the way.

The success or failure of Operation Overlord was not likely to be decided on 6 June 1944. The Allies made vast and elaborate preparations to ensure that D-Day would go well, but in the light of these it was overwhelmingly likely to do so. What mattered, in the days and weeks to come, was building up the Allied force and the area under its control faster than the Germans could assemble their reserves to confine it.

MONTGOMERY'S PLAN

All the Allied ground forces, British, American and Canadian, were under the command of General Montgomery's 21st Army Group. Probably more than any other general on the Allied side, he realized that if the Normandy invasion were to succeed, the Allies had to have a clear plan for the advance inland

Above: An American field artillery unit in action near the Normandy town of Carentan during the fighting in July 1944.

from the beaches. Montgomery developed such a plan and stuck to it so successfully that throughout the battle the Germans largely danced to his tune. His failure was that he never admitted when parts of the plan went astray. Thus, his relationship with other top Allied leaders was poor and his presentation of events to the press was unconvincing.

Montgomery's plan was for British and Canadian attacks on the Allied left to draw in the German reserves, while allowing the American units on the right to advance more rapidly.

Left: British troops on the watch for German snipers amid the rubble of Caen in late July.

Right: American infantrymen move cautiously past a burning Panther tank in a Normandy village.

This, in essence, is exactly what happened. However, it certainly did not all go "according to plan", though Montgomery claimed that it did.

The city and communications centre of Caen on the left (eastern) flank was meant to be captured on D-Day itself, but the advance from the beaches only got about halfway there. Caen did not fall completely until the third week of July, after several full-scale attacks by British Second Army, notably Operation Epsom at the end of June and Goodwood in July.

On the right flank the first task was to link the Utah and Omaha landings with the other beaches, which was achieved on 10 June. Next was an advance west across the Cotentin Peninsula and then north to capture Cherbourg, which was taken on 27 June. The port facilities, however, had been comprehensively wrecked by the Germans before their surrender. It would take many weeks to rebuild the port to full capacity, though it would land some cargoes during July.

GERMAN RESISTANCE

All the Allied forces were discovering just how tough the German armed forces were to beat, whether the best Army units like the Panzer Lehr Division or the teenage SS fanatics of the Hitler Youth Division. At all junior and middle-rank levels the standard of German leadership was higher overall, and the German troops fought in the knowledge that their tanks in particular were vastly better than anything the Allies had. However, the Allied generals were playing ever more solidly to their armies' strengths, in particular deploying a weight of artillery and air firepower to which the Germans could have no answer.

Through June into July US First Army was thus slowly and steadily gaining ground, while British Second Army was seemingly bogged down at Caen. Montgomery would have been delighted to have captured Caen sooner and pushed inland from there, but even without that his plan was working. Almost all the German Panzer divisions, the backbone of their force, were fighting on the British sector of the front. This prepared the ground for the decisive Allied breakout against the weaker part of the German line to the west of St-Lô in First Army's Operation Cobra from 25 July.

Below: US troops fighting near Périers during the advance across the Cotentin Peninsula.

Medium Bombers, 1942–5

By the end of the war the Allied air forces had long since gained total air superiority and this was nowhere better seen than in the operations of the thousands of Anglo-American medium bombers over Europe and in the Pacific.

The British and American medium bombers of 1942–5 include some of the most famous aircraft of the era. However, comparable Axis aircraft of the time remain little known. This is no coincidence. It simply reflects the failure of the Germans and Japanese to respond effectively to the Allied challenge in the air. (The Soviets also had few medium or heavy bombers but this derived from their total concentration on ground-attack aircraft and fighters, not a lack of effective air power.)

THE WOODEN WONDER

Britain only produced one aircraft in this class, but it was one of the best: the De Havilland

Below: A Mitsubishi Ki-67 *Hiryu* ("Flying Dragon") bomber, known to the Allies as "Peggy". In service from late 1944, the Ki-67 was fast (537kph/334mph) and better armed than previous Japanese bombers.

Mosquito. Almost 7,000 were built during the war, including photo-reconnaissance, ground-attack and night-fighter variants, as well as bombers.

All were fast and could carry a substantial load of bombs, guns or other equipment – and were made mainly of wood! In service with RAF Bomber Command, Mosquitos supported the main heavy-bomber force by flying diversionary raids and marking targets.

MAIN US TYPES

US medium-bomber aircraft included the Douglas A-20 Havoc/Boston, in service from early in the war, the North American B-25 Mitchell, Martin B-26 Marauder and Douglas A-26 Invader. A-designation aircraft were supposedly optimized for the (ground-) attack role, while the B-types were for the somewhat different medium-bomber mission. In practice

MOSQUITO B MARK 16

Mosquito bombers could carry a 4,000lb bomb (as shown) or a range of flares and target markers. Over 1,200 of the Mark 16 were built, the most numerous bomber version.

SPEED: 668kph (415mph)
RANGE: 2,400km (1,500 miles)
CREW: 2
ENGINES: 2 x Rolls-Royce Merlin 76/77; 1,710hp each
ARMAMENT: 1,800kg (4,000lb) bombs or flares; no guns

Right: A B-25H Mitchell in flight. This was one of the attack variants of the Mitchell, with a solid nose rather than a glazed bombardier's position. It carried eight forward-firing machine-guns and a 75mm cannon.

there was much overlap. The B-25 served throughout US involvement in the war and, like all the American aircraft mentioned above, was supplied in quantity to various Allies under Lend-Lease. Almost 10,000 were produced, in many variants that included several for the attack role, with up to 12 nose-mounted machine-guns and a 75mm (2.95in) cannon. Pure bomber versions could carry up to 2,700kg (6,000lb) of bombs.

There was little to choose between the B-25 and B-26 in performance or service career, though most of the 4,700 B-26s built for the USAAF were sent to Europe. The B-26 eventually had the lowest loss rate of any major USAAF combat aircraft, but when it was first introduced there were numerous accidents, avoided later by aircraft modifications and better pilot training.

Both the B-25 and B-26 saw combat from 1942, but the A-26 did not begin operations until mid-1944; over 1,000 were in use by 1945 and the type continued to serve for many years after the war. It carried similar armament to the Mitchell and Marauder but had only a three-man crew: pilot, navigator/bombardier and air gunner, who operated remotely controlled dorsal and ventral turrets.

GERMANY AND JAPAN
In the mid-war years German aircraft production planning was virtually non-existent, with a great deal of effort being wasted on minor upgrades of already obsolescent types and the development of numerous prototype designs. When this was set right in 1944–5, far too late, bomber production had to be virtually abandoned in favour of fighters. Thus the Dornier 217K, probably Germany's best bomber of the war, was taken out of production in late 1943.

One notable type that was introduced into service was the Heinkel 177. This unusual design appeared to be twin-engined but actually had a pair of engines in each wing to drive the single propellers. It had reasonable performance figures overall, but early examples in particular were very prone to disastrous engine fires.

Japan's aircraft industry never had the strength to compete with its enemies. Later-war bomber types included the Yokosuka P1Y "Frances", the Nakajima Ki-49 "Helen" and the Mitsubishi Ki-67 "Peggy". Fewer than 1,000 of each bomber were made.

MARTIN B-26 MARAUDER

First flown in 1940, the B-26 saw its first combat in the South Pacific in 1942. Most served in Europe, however. Some 5,300 were built in all and they were used by all the US services and various British Empire air forces.

SPEED: 462kph (287mph)
RANGE: 1,850km (1,150 miles)
CREW: 7
ENGINES: 2 x Pratt & Whitney R2800; 1,900hp each
ARMAMENT: 1,800kg (4,000lb) bombs; 12 x machine-guns

Resistance

Many people in occupied Europe judged that the defeat of Nazi Germany was a cause worth risking their lives for. Their bravery and sacrifice certainly played a large part in ensuring the ultimate Allied victory.

By its nature much resistance was a secret activity with few records kept. Many heroic acts no doubt also went unseen, so it is now impossible to quantify the effects of resistance in any meaningful way.

Resistance was certainly dangerous. Suspected resisters were routinely tortured and, if not executed, sent to concentration camps. Vicious mass reprisals against their communities were common. Unsurprisingly, therefore, more people took an active part in resistance as Germany's defeat became more certain, and those most active at the end were not necessarily those who had the best record throughout.

Resistance took many forms and might not even have had any long-term effect on the war.

Above: Resistance fighters of the Polish Home Army in action during the Warsaw Rising of 1944.

Norwegian history teachers, for example, refused to teach the Nazi-approved syllabus and stood firm to win their point, despite some being deported to concentration camps. Other more significant forms of quiet heroism came from those who hid Jews from the SS, many thousands in Poland and the Netherlands alone.

Some non-violent methods had clear military results. French railway workers deliberately doing their jobs slowly and incompetently may have done more to hinder the movement of German troops and supplies in France in 1944 than the Allied bombers or the active sabotage efforts of their resistance comrades. Similarly, up to 2,000 Anglo-American aircrew who had been shot down were helped to escape from German-occupied territory, an important increment to Allied strength.

ALLIED ASSISTANCE

The Western Allies had various organizations that supported resistance. The USA's Office of Strategic Services (OSS) was involved both in intelligence-gathering (clearly an important function of resistance generally) and the support of military resistance. Britain split these functions between the secret service, MI6, and the wartime Special Operations Executive (SOE). Britain's MI9 and the American MIS-X helped escapees and those who assisted

Left: Resisters in newly liberated Marseille show off their weapons. Resistance was one area in which female combatants played a full part.

them. SOE and OSS trained many operatives for covert service in the occupied countries and had extensive technical sections designing and making special weapons, forging identity documents and carrying out many similar tasks.

HORROR IN YUGOSLAVIA

Resistance fighting was most intense of all in Yugoslavia. At the end of the war the Partisan movement, led by Josip Broz Tito and by then backed by both the Soviets and the West, effectively liberated the country from the Germans after four years of particularly bloody struggle. The Partisans were the victors after what was really a civil war between them, the Četniks (mainly Serbian nationalists) and the Ustašas (mostly Croatian fascists). Each fought the others and with and against the occupying Germans and Italians at different times. Most of the 1 million plus Yugoslavs who died during the war were killed by other Yugoslavs.

The situation in Western Europe was never as extreme, as the example of France shows. There was certainly tension and distrust between communist and non-communist resistance groups, but almost the whole resistance movement agreed in 1943 to work together under the command of the Free French leaders in exile. In 1944 there were extensive and successful plans to co-ordinate resistance

Right: Josip Broz Tito (nearest camera), leader of the Yugoslav Partisan movement, meeting "Draža" Mihailović (dark hair in front row) of the rival Četnik group, which initially received most Allied support.

operations with the Normandy invasion. Many resisters were young men who had taken to the hills and forests to avoid compulsory work service in Germany. These Maquis groups played a notably important part in assisting the Allied invasion

Above: General Charles de Gaulle, leader of the Free French, making a broadcast to his countrymen.

of southern France. Perhaps 100,000 French people died in resistance activities or in German reprisals against them.

Collaboration

Collaborators had many motives. Some genuinely supported Nazi ideas; some were merely trying to survive as best they could in impossible circumstances; and some thought of themselves as patriots honestly trying to serve their countries.

Who were collaborators? Government ministers in occupied countries who decided in the early-war period to work with the Germans could be regarded as collaborators – or were they just being realistic in making the best arrangements they could with the country that had effectively won the war? Industrialists might run factories making goods for the Germans and money for themselves, but they were also keeping their workers in decent jobs and preventing them being deported as slave labour to Germany. A village mayor or a policeman had no choice but to obey some German orders, but when did he go beyond that into enthusiastic help? And, at the simplest level of all, was a

Above: A Waffen-SS poster seeks Norwegian recruits to fight against communism. Norwegians and Danes were eagerly enlisted by the SS.

woman in a relationship with a German soldier a shameful traitor or in fact doing something all too human with a young man scared and far from his home? It was not always easy to decide.

Almost every country had native fascists and anti-semites who worked willingly with the Nazis, genuinely believing that they were acting patriotically. For example, Anton Mussert's party in the Netherlands had 50,000 members. But he, and his like elsewhere, were given no real power by the Germans.

Poland was probably the country that saw the least collaboration with the Germans, but even there another kind of collaboration was seen. Many pro-communist Poles helped the Soviets consolidate their control from 1944, and would have been seen as collaborators by most of their compatriots.

JOINING THE SS

Putting on a German uniform and fighting for the Germans was certainly collaboration in most people's eyes. In fact many people born outside Germany enlisted in the Nazi Party's armed elite, the Waffen-SS. They included both ethnic Germans recruited from abroad – *Volksdeutsche* (as the Nazis

Left: Women who had relationships with Germans being forced to parade through the streets of Paris with shaved heads and in their underclothes after the liberation.

called them) – and foreigners from "Germanic" countries like Norway or Denmark. Later in the war the net was cast wider; Ukrainians, Croatian Muslims, Latvians, Estonians and others were all accepted. Eventually, over half of Waffen-SS soldiers were not native Germans.

Substantial numbers of Soviet citizens ended up working for, and sometimes fighting alongside, the German Army. By 1942 many of Germany's units included significant numbers of auxiliaries – *Hilfswillige* (or "Hiwis"). Many of these were former prisoners of war who had made this choice to escape the privations and likely death of captivity. At first they worked mainly in non-combatant support roles, as drivers, cooks and the like, but later they became fighting troops. In Normandy in 1944 German Army regiments regularly included an *Ostbataillon* (or "East Battalion") of such men and some fought effectively.

RUSSIAN NATIONALISTS

Other units that were formed by the Germans attempted to use anti-Soviet feeling and even appeal to Russian nationalism. These included various units of non-Slav Soviet citizens – Cossacks, Armenians, Georgians and others – who often fought in anti-partisan operations. Russians as such were also involved. A Russian National

Army of Liberation was formed in the Bryansk region. Better known as the Kaminsky Brigade, it gained a brutal reputation, notably during the Warsaw Rising of 1944.

A senior Red Army officer, General Andrei Vlasov, was captured in 1942 and agreed to raise a substantial anti-Stalinist Russian force. Top-level German support was half-hearted and only two divisions were formed,

but not used. In the last days of the war one of these turned against the Germans and fought alongside the Czech resistance during the battle for Prague.

The post-war fate of identified collaborators was often harsh. Millions of Soviet citizens were turned over to the Soviet authorities by the Western Allies. Most ended up in the GULag at best; Vlasov and his officers were executed.

Crushing Soviet Victories, 1944

The summer battles of 1944 began with what Soviet historians call the "destruction of Army Group Centre", and by the end of the year Romania and Bulgaria were fighting alongside the Soviets with the front line well into Poland and Hungary.

Misled by the Soviet advance into Romania in the spring of 1944, Hitler and his generals expected the Soviets to continue to concentrate on the southern part of the front for their next big attacks. Accordingly they transferred most of the tanks and many other heavy weapons from German Army Group Centre in Belorussia, just at the time when the Soviets were concentrating over 2 million troops and vast resources to attack it.

ADVANCE IN BELORUSSIA

The attacks began on 23 June and, partly because of the unimaginative command of Field Marshal Ernst Busch (who was quickly replaced), soon smashed the German front. Three German armies – Third Panzer, Fourth and Ninth – each lost the majority of their

Above: A horse-drawn Soviet 45mm (1.77in) anti-tank unit advancing into Poland in 1944.

strength, around 400,000 men in all. Minsk, the offensive's initial target, was captured on 11 July.

To the north Soviet attacks beginning on 4 July cleared the Germans out of much of Latvia and Lithuania by the end of August. By November virtually all the Baltic States were in Soviet hands, except for the

Courland Peninsula, where some 20 German divisions would hold out until the end of the war. Finland, allied with Germany since 1941, had also been attacked in June; and the Finns then agreed an armistice with the Soviets in September.

During July the Soviet Belorussian fronts continued their advances at pace and were joined by new attacks by the Ukrainian fronts to the south. Brest-Litovsk and Lvov were both taken late in the month. By 15 August Soviet troops were on the east bank of the Vistula River opposite Warsaw.

WARSAW UPRISING

Resistance fighters of the Polish Home Army had already been helping the Soviet advance and on 1 August they began an all-out uprising in Warsaw. They fought on until early October, by which time over 200,000 Poles had been killed.

The Soviets made no attempt to help them, claiming (not entirely implausibly) that they needed to resupply their front-line forces after the recent advances before they tried to cross the Vistula. However, it also suited Stalin's grim purpose to have the Germans kill off anti-communist Polish patriots for him. Similarly, when Slovaks rose against the Germans in late

Left: Red Army infantry moving up to the front for the Soviet attack on Lvov in July 1944.

Above: Soviet infantry assault pass a burning Panther tank, eastern Poland, summer 1944.

August, the Soviets did little to help. This rising was also largely put down by late October.

INTO THE BALKANS

With its northern forces halting on the Vistula, the Red Army began new attacks in the far south on 20 August. Within days much of German Sixth Army was destroyed near Kishinev. King Michael of Romania mounted a coup on 23 August and his new government brought the country (and its large army) into the war on the Soviet side.

As the German forces continued their fighting retreat through Romania and into Hungary, Bulgaria was next to fall. The last in a sequence of government changes brought a partly communist group to power on 9 September, the day after Soviet troops had crossed into Bulgaria from Romania. The Bulgarians, too, would now join the Allied forces on the Eastern Front, while some Soviet troops remained in their country to ensure communist control.

With these threatening advances to their rear, German troops began pulling out of Greece and Yugoslavia. Yugoslav Partisans liberated Belgrade on 19 October and by the end of the year controlled roughly half of their country. British troops from Italy arrived in Greece to take over there in October too.

By December the Germans had again managed to build a defensive front. The principal fighting was in Hungary, where Budapest was surrounded on 26 December and would endure a bitter siege until mid-February.

Marshal Zhukov (1896–1974) was the leading Soviet soldier of the war. His first big achievement was the decisive defeat of the Japanese at Khalkhin Gol in 1939. In 1941–2 he was repeatedly switched between staff jobs and front commands by Stalin. He was credited with masterminding the defence of Leningrad, Moscow and Stalingrad, although in each case the timing of his arrival was fortuitous. Zhukov then directed various of the Red Army advances up to and including the capture of Berlin in 1945. As a commander he was ruthless and thorough, one of many such leading generals who made the shambolic Red Army of 1941 into the world's most feared fighting force by 1945.

Above: Marshal Zhukov during the advance to Berlin in 1945.

Left: An SU-76 assault gun in action in a village in south-eastern Europe during the Soviet advance in the late summer of 1944.

Heavy Tanks, 1942–5

Among the most fearsome weapons of the war, heavy tanks were equally formidable in action against enemy strong points and armour. With their great gun power some could pick off opposing tanks at well over a kilometre.

At the start of the war the main users of heavy tanks were Britain and the USSR. The British nomenclature "infantry tanks" well described their function for both nations, which was to accompany and support the infantry assault. The far more formidable machines in service in the second half of the war retained this role in part, but their anti-armour performance generally became more important. The tanks included here normally served in separate heavy tank or infantry support units.

Britain's last infantry tank was the Churchill, in service in a range of variants from 1941. The initial version was rushed into service and, as well as being woefully under-gunned (with a 2pdr/40mm), was extremely unreliable at first. Later marks had the 6pdr (57mm), then a 75mm (2.95in). All versions were well-armoured but slow, though they were very good at climbing slopes and crossing obstacles. Over 6,500 of all marks were built, plus several hundred more for specialized engineer tasks.

SOVIET POWER

The performance of the Soviet KV-1 heavy tanks in service in 1941 came as a shock to the Germans but gradually improved German guns meant that they lost their invulnerability later, while their own 76.2mm (3in) weapon was inadequate against newer German tank types. Small numbers of a stop-gap KV-85, with an 85mm (3.35in) gun, were produced in the second half of 1943 before production ended. Some 10,000 of all models were built.

Their replacement was the Josef Stalin series. A few examples of this design were built, each with 85mm and 100mm (3.94in) weapons before a final decision on the 122mm (4.8in) D25 gun was made. This had a slightly poorer armour-piercing performance than the 100mm, but this was compensated for in part by its more powerful high-explosive shell for the other aspects of the tank's role. One definite disadvantage, however, was the

Below: Tiger tanks preparing for an attack on the Eastern Front.

JOSEF STALIN 2

The JS-2 was deliberately kept relatively small – it was no heavier than the KV or the German Panther. Even so, it packed a fearsome punch and had well-sloped armour for effective protection.

WEIGHT: 44.7 tonnes
LENGTH: 8.33m (27ft 4in)
HEIGHT: 2.72m (8ft 11in)
WIDTH: 3.12m (10ft 3in)
ARMOUR: 120mm (4.7in)
ROAD SPEED: 37kph (23mph)
ARMAMENT: 1 x 122mm (4.8in)
D25T + 4 x machine-guns

TIGER 2/KÖNIGSTIGER

For all their great fighting power, only 454 Tiger 2s were built, serving in only a handful of heavy-tank battalions. The group shown were photographed for a propaganda film in France in summer 1944.

WEIGHT: 69.4 tonnes
LENGTH: 7.23m (23ft 9in)
HEIGHT: 3.07m (10ft 1in)
WIDTH: 3.73m (12ft 3in)
ARMOUR: 180mm (7.1in)
ROAD SPEED: 35kph (22mph)
ARMAMENT: 1 x 8.8cm (3.46in)
    KwK 42 + 2 x machine-guns

Above: M26 Pershings of the US 2nd Armored Division in the streets of Magdeburg on the Elbe in 1945.

slow rate of fire of the 122mm and the fact that only 28 rounds could be carried. Probably about 4,000 of the JS-1 and JS-2 variants saw service; a JS-3 version was in production by 1945 but did not see combat.

TIGERS IN ACTION

The first of the German designs that outmatched the KV-1 was the famous PzKpfw 6 Tiger, introduced in September 1942 on the Eastern Front. Design work on the Tiger had begun before Operation Barbarossa revealed the strength of the Soviet armour but was then greatly speeded up. When introduced, the Tiger mounted the then most powerful tank gun in the world (the 8.8cm/3.46in KwK 36) behind very thick armour; it quickly gained a fearsome reputation for its gun power and defensive strength. Mobility and reliability were never strong points, but it was still a tough opponent when the last of 1,355 came off the production line in August 1944.

Its replacement was the Tiger 2 (also know as the Königstiger; 454 built by 1945). It first saw service in Normandy in July 1944. At some 70 tonnes it was massively armoured and its 8.8cm KwK 42 was substantially more powerful than the Tiger 1's gun. Predictably its weak point was its mobility, but in the defensive battles then being mainly fought by the German Army this was not necessarily a great disadvantage.

After delays in the mid-war period, the modest capabilities of the M4 Sherman, in comparison to the German Tigers and Panthers, finally spurred production of something better for the US Army from later in 1944. The M26 Pershing first saw combat in February 1945 and a few hundred were shipped to Europe by VE-Day. It carried an effective 90mm (3.54in) M3 gun and had a good balance of protection and engine power.

Totalitarian Rule: Germany and the USSR

For all that the two countries were the bitterest enemies, there were many close comparisons in the methods of government in Nazi Germany and the USSR, starting with the cruel and murderous nature of the regimes.

It is hardly surprising that there were a great many similarities between Nazi Germany and the Communist USSR in the WWII period. Both countries were dominated by a single political party whose organs and whose secret police featured in every corner of national life. And at their head was a single ruler with the power of life and death and absolute control over every aspect of state policy.

POLICE AND SECURITY

Both countries' regimes had a unified apparatus for security, policing and public discipline matters. In Germany the Reich Main Security Office (RSHA), a major division of Heinrich

Above: Lavrenti Beria controlled the USSR's security apparatus from 1938–53, murdering at Stalin's command and for his own pleasure.

Below: One of the 1944 Bomb Plotters faces the ranting Nazi judge Roland Freisler at his show trial before his inevitable execution.

Himmler's SS, was the controlling body. Its main chiefs were first Reinhard Heydrich and later Ernst Kaltenbrunner; its subordinate departments included the SS security service and the Gestapo secret police. The RSHA was surprisingly small, only some 50,000 people in all in 1944, but it also maintained an extensive network of informers all too ready to denounce "defeatists" or "asocials".

In the USSR the organization varied at different times in the war but included both the NKVD and NKGB. However, both came under the control of Lavrenti Beria, the Commissar for State Security. Unlike the rather squeamish Himmler, who disliked personal involvement

in the atrocities he ordered, Beria was himself a vicious and enthusiastic torturer and killer.

The principal instruments of coercion in both countries were the prison camps. No one knows how many unfortunates suffered in the Soviet camps, the GULag, but it is clear that to be sent to the camps was virtually a death sentence. Estimates suggest that during the war 1 million prisoners a year died of overwork, starvation and every other kind of cruelty. The process of selection was haphazard; many prisoners had done nothing to offend the regime.

Most of those in Germany's concentration camps were in some way obnoxious to the Nazis: habitual criminals, left-wingers and many more. At first prisoners were at least meant to be kept alive to work (the death camps where Jews were sent had a different mission). Camp prisoners slaved in vile conditions in quarries, mines and factories – many run as highly profitable businesses by the SS. Half a million is a common estimate of the number who died in the concentration camps.

Discipline in the armed forces was also stunningly harsh. The German Army executed at least 10,000 of its men (and probably many more) for military crimes like cowardice and desertion. (The US Army executed one.) The Red Army had numerous punishment units to which several hundred thousand men were sent for astonishingly trivial reasons. These units were used on repeated near-suicidal missions, which few of their soldiers survived.

THE HOME FRONTS

Generally speaking civilian life was relatively comfortable for Germans until 1943 at least. Rations were reasonable; consumer goods were available; women were not required to do war work; and many prosperous families even had servants.

None of this was true in the USSR. Only workers in important industries got rations much above starvation levels. Millions of people were moved to the new industrial and mining areas out of the Germans' reach. Employees in the new factories built in 1941–2 often started work before the roof was even on

Below: Prisoners at the Nazi Sachsenhausen concentration camp in 1938. About 100,000 died there.

and slept among the machines because they had no houses. Remarkably they did so very willingly on the whole, a level of commitment that was central to the USSR's victory in what its leaders taught the people to call the Great Patriotic War.

Indeed despite all the cruelties of the ruling regimes, the German and Soviet peoples fought hard for their countries throughout the war.

Right: A German poster appealing for donations of old clothes and woollens for the war effort.

ASSASSINATING HITLER

There were several attempts to assassinate Hitler. The most nearly successful was on 20 July 1944, when he was injured by a bomb placed in his conference room. The group of officers responsible planned a military coup, but this fell apart when it became clear Hitler had survived. About 5,000 people were executed in the aftermath. Hitler became even more mentally unstable than before and trusted his generals even less.

The Defeat of the Luftwaffe

American air victories over Germany in early 1944 set both the British and American bomber forces free to smash Germany's war effort. Heavily outnumbered and practically out of fuel, the Luftwaffe could do little to resist.

Two vital command decisions ensured the crushing defeat of the Luftwaffe in early 1944. Although the leaders of the American bomber force had not fully appreciated the importance of fighter escorts for their attacks, in late October 1943 General "Hap" Arnold, Chief of Staff of the USAAF, ordered that all P-51 Mustang production for the next three months was to go to Eighth AF for bomber support. The second decision came in January 1944 when General James Doolittle, new commander of Eighth AF, ordered his fighters not to be tied to the close-escort mission but to seek out and destroy their German counterparts. Together these set the scene for a series of massive air battles in which the USAAF would win air superiority over Germany.

Early 1944 saw RAF Bomber Command continuing to lose heavily in the Battle of Berlin. This culminated in a disastrous attack on Nuremberg on the night of 30–31 March, in which 96 out of 795 bombers were lost. Although Air Marshal Harris bitterly opposed the decision and made as many "area bombing" attacks on German targets as he could get away with, it was fortunate for his men that they were mainly sent against targets in France, over the next several months, in support of the coming D-Day landings.

"BIG WEEK"

The first real chance for the American forces to put the new policy into effect came in mid-February, in a series of attacks on the German aircraft industry known as "Big Week". Contrary to what the Allies believed at the time, these barely dented German aircraft production, which continued to increase until September 1944 because of improved organization. Nor was there a clear-cut victory for either side in the air – the American forces lost some 250 bombers in return for 355

Left: A reconnaissance photo of bomb craters around the Dortmund–Ems Canal. Once used to transport U-boat parts, the canal was emptied by RAF attacks in late 1944.

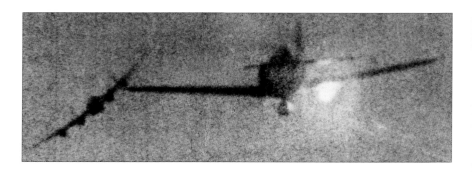

Above: A camera-gun picture from a British fighter shooting down an Fw 190 trying to attack a Lancaster bomber over Germany in late 1944.

German fighters downed. However, over 100 irreplaceable German pilots were killed. Fighter-pilot training and experience levels on the American side were on the increase; other than for a diminishing band of top aces, the reverse was now true for the Germans.

The process continued relentlessly, notably with missions to Berlin itself in March. Even though the US heavy bombers, like RAF Bomber Command, came under General Eisenhower's control on 1 April in preparation for D-Day, they continued to make some telling attacks on targets in Germany and Central Europe.

OIL ATTACKS

From May both the USAAF and RAF Bomber Command made an increasing number of strikes on Germany's synthetic oil production sites. Although Harris had been right to deride previous attempts to pick off key German target systems as ineffective "panaceas", this proved to be something different. By September 1944 German aviation fuel production was around 7 per cent of the already inadequate level achieved in May.

This hampered operations and made training of the new pilots now in urgent demand virtually impossible. Those who did leave the training system were increasingly easy meat for the better prepared Allied fighters.

In August, in perhaps the clearest token of how things had changed, RAF Bomber Command returned to daylight operations over Germany. From then until attacks were wound down in the final weeks of the war, both British and American heavy bombers ranged almost at will over Germany, flattening cities and devastating industries with a weight of destructive power and levels of accuracy out of all proportion to anything ever before achieved.

Below: A damaged B-17 Flying Fortress drops its bombs on Berlin during a raid in August 1944.

DRESDEN

Since the war, and to some degree during it, some people have questioned the morality of heavy-bomber attacks on enemy populations and have doubted whether they were an efficient use of resources. The bombing by the RAF of the city of Dresden on 13–14 February 1945 has been particularly controversial. There were no military targets of importance in the city and it was packed with refugees, yet it was smashed by over 700 bombers and as many as 70,000 or perhaps more people were killed.

Above: Bodies litter the streets of Dresden after the attacks.

Fighters, 1943–5

By late 1944, thanks to the British, American and Soviet fighter forces and their range of superb aircraft, Allied heavy bombers ranged freely over Germany and Japan and ground-attack aircraft dominated the Axis land forces on all fronts.

From mid-1943 Allied air superiority over every battlefront was clear, and during 1944 was extended to cover both the German and the Japanese homelands by day and by night. It was the Allied fighters that made this happen. Although Germany and Japan both continued to field highly effective piston-engined fighters, overall the Allies had a qualitative advantage (other than over Germany's jets) to add to their enormous numerical superiority.

ALLIED DESIGNS

The USAAF was the world's strongest air force in 1944–5 and had the fighters to match. The main types were the North American P-51 Mustang and the Republic P-47 Thunderbolt, both dating from 1940–1, although the earlier P-38 Lightning also remained in use.

The P-47 seemed the more promising design initially. It was based around a particularly large and powerful radial engine, which gave it an impressive rate of climb and dive despite its imposing size and weight. The P-47D version was the most built subtype of any fighter ever, with over 12,000 made.

After an unpromising start the P-51 Mustang developed into the aircraft that did more than any other to win the air war over western Europe. Fitted with the initial Allison engine, the Mustang had disappointing flight performance, especially at

P-51D MUSTANG

The P-51D is regarded as the definitive wartime version of the Mustang. Like its B and C model predecessors, it was fitted with an American-built Packard Merlin engine but had two extra machine-guns and a "bubble" cockpit canopy for improved pilot visibility. Over 8,000 P-51Ds and some 16,000 of all P-51 marks were built.

SPEED: 703kph (437mph)
RANGE: 2,655km (1,650 miles), with drop tanks
CREW: 1
ENGINE: Packard Merlin V-1650; 1,695hp
ARMAMENT: 6 x 0.5in (12.7mm) machine-guns

Left: Late-war P-47 Thunderbolts were fitted with a "bubble" cockpit canopy to improve pilot visibility.

altitude; but in 1942 a version with a Rolls-Royce Merlin engine was tested and the Mustang was transformed. Fitted with drop tanks for additional fuel, the new aircraft had the range to escort bombers from England to Berlin and beyond and was superior to almost all Luftwaffe fighters when it got there, a combination of range and performance previously thought to be impossible.

In the final stages of the war many British fighter units continued to use the Spitfire Mark 9, introduced in 1942. This was joined by variants, notably the Mark 14, in which the Merlin engine was replaced by a more powerful Rolls-Royce Griffon. These gave nothing away to contemporaries in speed and manoeuvrability but, like earlier Spitfires, lacked range.

Completing the formidable Allied line-up was a range of impressive Soviet designs, the best of which came from the Yakovlev and Lavochkin design bureaux. The Yak-9, introduced in 1942, was an effective upgrade of the earlier Yak-7, with a more powerful engine and better armament fit. Its upgrade to the 9U version in 1944 was even more impressive. The ultimate Yak fighter of the war, also introduced in 1944, was the confusingly designated Yak-3.

Above: A formation of Yak-9D escort fighters in flight. Over 16,500 Yak-9 fighters were built.

This was an extremely fast and manoeuvrable design, probably the best fighter of the war on the Eastern Front. The La-5 and La-7, developed from the previous LaGG designs, were both also very effective if not quite the equals of the Yak types.

AXIS REPLIES

Germany's Focke-Wulf 190 had outclassed the RAF's Spitfires when introduced in 1941. Versions of this design and later marks of the Messerschmitt Bf 109 remained the principal German fighters until the end of the war. By 1944 the Bf 109 and the radial-engined versions of the Fw 190 struggled to compete with the best Allied aircraft. The Fw 190D, fitted with an in-line engine and available from late 1944, was faster and more formidable.

In the Pacific War Japan's aircraft industry also failed to keep up with new designs. As a token

of this, the Kawasaki Ki-61 "Tony" (one of the most significant later-war Japanese fighters) was only one of several Japanese aircraft relying on licence-built versions of German engines. The best Japanese Army fighter of the war was the Nakajima Ki-84 "Frank", in service from April 1944.

Below: A Griffon-engined Spitfire Mark 14 in flight. The Mark 14 had a top speed of 721kph (448mph).

FOCKE-WULF 190

Over 20,000 Fw 190s were built in numerous variants, including the radically different D model and successors with an in-line rather than a radial engine. The Fw 190 served as a fighter, including some with heavy armour and armament for the bomber-attack role, and as a fighter-bomber.

SPECIFICATION: Fw 190A-3
SPEED: 640kph (398mph)
RANGE: 800km (500 miles)
CREW: 1
ENGINE: BMW 801 D-2; 1,730hp
ARMAMENT: 2 x 20mm (0.79in) cannon + 2 x 7.92mm (0.312in) machine-guns

Aircraft Ordnance and Electronics

Air attacks were meaningless without effective weapons and systems to guide aircraft to their targets. Bombs, rockets, radar, navigation aids and equipment to jam or home in on enemy transmissions all saw substantial development during the war.

It is well known that WWII saw much behind-the-scenes innovation in radar and other aspects of aircraft electronics, but developments in seemingly mundane fields like bomb design were also significant.

BOMBS AND ROCKETS

Many different bombs were used, from 1.8kg (4lb) incendiaries, dropped by the millions, to the RAF's 9,980kg (22,000lb) Grand Slam of which 41 were used in 1945. By the later stages of the war a typical British heavy-bomber load for an attack on a German city would include a single "cookie" (a large, high-capacity, high-explosive bomb) to blast buildings open, as well as a range of incendiaries to set them on fire. In the 1945 US attacks on Japanese cities incendiary bombs were the principal types used. By 1945 ground-attack and anti-shipping aircraft in US, British and Soviet service were also using un-guided air-to-ground rockets as a major part of their armoury.

Although most bombs were "dumb" free-fall weapons, various guided and powered bombs

Above: A *Mistel* composite aircraft. The pair would be flown to the target by the fighter pilot who would then release the crewless but explosive-filled bomber to complete the attack. *Mistel* aircraft achieved little success.

were developed. German types included the rocket-powered Hs 293 and the free-fall Fritz-X; the American Azon type was similar to the Fritz-X. All were radio-controlled from the dropping aircraft, which had to keep the bomb and the target in sight throughout, always tricky and often dangerous. The American Bat type was potentially more capable because it included its

own radar set, which it used to home in on its target. German successes included the sinking of the Italian battleship *Roma* by a Fritz-X in September 1943.

ELECTRONIC WARFARE

All major nations had some degree of radar capability in 1939, but Britain, the USA and Germany were the only ones to make substantial developments in this field. The accuracy of radar sets increased substantially during the war, as did the variety of methods to fool them and in turn the devices made to overcome these.

The most used anti-radar device was strips of metal foil dropped from an aircraft to give a mass of false radar returns on enemy screens – provided the strips were cut to the correct size according to the radar wavelength. This was first used by

Left: A Messerschmitt 110 night fighter fitted with Lichtenstein radar.

Right: Ground crewmen loading a 4,000lb (1,815kg) High Capacity bomb into a Lancaster.

British forces in July 1943 and in updated forms provided significant protection to British and American aircraft over Germany and Japan to the end of the war.

The Luftwaffe was the only air force in 1939 to appreciate the difficulty of finding bombing targets by night or in bad weather and had various radio-beam systems (Knickebein, X-Gerät and Y-Gerät) in service to achieve this. These were used with some success during the Blitz in 1940–1, but once British scientists had worked out how they operated, the systems were relatively easy to jam.

The British Gee, introduced in 1942, worked on a similar tri-angulation principle and was also jammed after a few months' use. The later Oboe used radar to measure the range between the aircraft and various ground stations. A disadvantage of all these was that their range was limited by the curvature of the Earth to about 450km (280 miles).

Slightly later still was the H2S system (used by American forces as H2X and in an improved APS-20 form), a ground-mapping radar operated by the aircraft independently of any ground station. This worked best when the ground below had distinctive features like a river or coastline. With devices like these, the RAF heavy-bomber force of 1944–5 was able to attack at least as accurately at night as the American bomber groups by day.

All such devices brought a disadvantage, however. The enemy could use any radio or

Right: Ground crewmen loading a 4,000lb (1,815kg) High Capacity bomb into a Lancaster.

radar transmitter in an aircraft to home in on it. Late-war British Mosquito night fighters, for example, had effective equipment called Perfectos and Serrate to do just this to their German counterparts.

Right: An Avro Lancaster in flight with the distinctive bulge of its H2S ground-mapping radar clearly visible under the rear fuselage.

From Normandy to the Rhine

At the end of July 1944 the continued Allied pressure in Normandy finally cracked the German front. Within a month the Allied troops had liberated most of France and Belgium but hopes of winning the war in 1944 would soon be abandoned.

On 25 July US First Army was able to begin Operation Cobra to break out from the Normandy bridgehead. Over the next few days the attack made good progress south past Avranches. By 1 August the German front in the area had crumbled away and a new Allied formation, US Third Army led by General George Patton, burst out into the open. Part of Third Army cleared Brittany to the west, eventually taking the ports of St-Malo and Brest, but other units headed south-east for Le Mans and the interior of France.

Hitler now intervened. Although all the other Allied forces were also pushing foward and his generals were sensibly beginning to think about a retreat, he ordered an attack through Mortain to Avranches to cut off the Third Army advance where it began. This attack was fought to a standstill on 7–8 August and by then the whole German force in Normandy was in deep trouble.

THE FALAISE POCKET

Around 20 divisions were caught in a pocket between British and Canadians in the north and Americans in the south. They had no option but to retreat frantically to the east past Falaise, while trying to fend off Allied attacks all the way. Although many of the German troops in the so-called Falaise Pocket did manage to escape before its neck was closed, they no longer formed effective military formations. And, to complete the Allied victory in Normandy, Allied troops were already across the River Seine west of Paris.

There was yet more bad news for the Germans. On 15 August French and American troops landed on the Mediterranean coast between Cannes and Toulon. The German forces in southern France were too weak to provide real opposition and were also being seriously harassed by French resistance fighters. By the end of August the invasion force was advancing steadily up the Rhône Valley

Map legend:
— Front line 13 June 1944
– – Front line 18 July 1944
······ Front line 13 Aug 1944
—— Front line 26 Aug 1944
– – Front line 14 Sept 1944
······ Front line 15 Dec 1944
■ German-held until May 1945

LIBERATION OF FRANCE
After a long and bitter struggle in Normandy, Allied forces liberated the rest of France in only a few weeks.

Above: Allied Sherman tanks crossing the Nijmegen Bridge on their way to Arnhem.

Above: Many German troops and vehicles were hit by Allied air attacks in the Falaise Pocket.

and would link up with troops advancing from Normandy on 12 September.

After the slow slog through Normandy the campaign was transformed. Paris was liberated on 25 August. By the end of the month most of France was free and most of Belgium followed in the next few days. By now, the Allied forces were out-running their supply resources and the top generals were arguing over what to do next. The port of Antwerp, which had the capacity to land all the supplies the armies were likely to need, had fallen on 4 September, but its estuary was still in German hands. The best plan would probably have been to have immediately mounted a campaign to clear the estuary and solve the supply problems, but this was not done until October and the first cargoes were not landed until late November.

A BRIDGE TOO FAR

Instead Montgomery persuaded Eisenhower, the Supreme Commander, that the best plan was to cross the series of river barriers blocking the route into Germany in a joint airborne and ground-forces attack. Operation Market Garden seized several major bridges but the last, over the Rhine at Arnhem, could not be taken. The Allied advance ground to a halt all along the front by the end of September.

While the British and Canadian forces were finally clearing the Scheldt estuary, the American armies were fighting hard for small gains around Aachen and in the Saar and Moselle regions to the south. It was an autumn and early winter of bitter disappointment after the high hopes of only weeks before. And the Germans were recovering fast from their Normandy disaster and assembling reserves for a new attack.

GENERAL GEORGE PATTON

Patton (1885–1945) was probably the most able American general of the war, though his talents were marred by showmanship and a love of publicity. He successfully commanded US troops in North Africa and Sicily but was then sidelined for a time after striking two hospitalized soldiers he wrongly accused of malingering. He commanded US Third Army in 1944–5, leading it particularly brilliantly during the advance in August 1944 and in the Battle of the Bulge in December.

Right: General Patton, photographed shortly before the Sicilian campaign during which he quarrelled fiercely with General Montgomery.

Self-propelled Artillery

In a fast-moving war, armoured units needed their artillery to be as mobile as their tanks and a variety of self-propelled guns was developed to fill this role. Most were based on standard field gun weapons but some were in the heavy artillery class.

Weapons in this category are those self-propelled (SP) guns whose armament was designed principally or even exclusively for use with high-explosive or similar ammunition – and which normally lacked the armour protection to allow their regular use within direct-fire range of the enemy.

Designs in this class in both British and American service were regarded as pure artillery weapons, serving in the American case in "armored field artillery regiments" and operating like towed field artillery units with additional mobility. For the Soviet and German Armies the distinction between weapons of this type and those described as assault guns is not always clear-cut.

THE WESTERN ALLIES
The earliest Anglo-American types made their debuts in the Desert War in 1942. The first British design was the Bishop, a 25pdr (87mm) gun mounted in a high box-like superstructure on a Valentine tank chassis. This was not a success – among other faults the mounting greatly limited the gun's elevation, halving its normal range.

The most-produced US design first fought at Alamein in 1942 in British hands. This was the M7 105mm HMC (or Priest in British service). It carried a standard 105mm (4.13in) M2 Howitzer (with restricted elevation, though not as badly as the Bishop) on a chassis based on the M3 Lee tank. It continued in use through 1945. British forces used a similar vehicle, fitted with a full-elevation 25pdr and known as the Sexton.

Other American designs included two light types, mounting 75mm (2.95in) guns. The M3 GMC was mounted on a converted M3 half track and the M8 HMC on a modified M5 Stuart tank chassis. Neither was a great success.

A small number of M12 155mm GMCs, carrying a WWI-era 155mm (6.1in) gun, were used in north-west Europe in 1944–5. Other designs using rather more modern 155mm guns and howitzers were developed but not produced in time to see action.

Below: A German Wespe 10.5cm battery with vehicles whitewashed for winter camouflage.

BISHOP

The need to produce artillery weapons that could keep up with fast-moving tank battles in the Desert War led to the rushed development of the Bishop. About 100 saw action in North Africa, like this one, shown in Tunisia in 1943.

WEIGHT: 17.4 tonnes
LENGTH: 5.53m (18ft 2in)
HEIGHT: 2.76m (9ft 1in)
WIDTH: 2.61m (8ft 7in)
ARMOUR: 60mm (2.4in) max.
ROAD SPEED: 24kph (15mph)
ARMAMENT: 1 x 25pdr (87mm) Gun Mark 2; 32 rounds carried

Right: M7 105mm guns in front of Notre Dame during the French 2nd Armoured Division's triumphant liberation of Paris in August 1944.

THE EASTERN FRONT

Soviet designs covered here include several mounting versions of the 152mm (5.98in) gun. The early-war KV-1 heavy tank was accompanied by a 152mm-armed KV-2. This was well-armoured, but its high, boxy shape made it clumsy and vulnerable. The next type, the SU-152, also used the KV chassis but was much more effective, in service from the Battle of Kursk in 1943. Around 700 were built. It was replaced in production in 1944 by the JSU-152, carrying the same weapon but based on the JS-series of heavy tanks. All of these weapons also had a significant anti-tank capability.

German designs were more varied. The Wespe was roughly equivalent to the American M7.

It carried the standard 10.5cm leFh 18 howitzer on a Panzer 2 chassis and was allocated to the artillery units of Panzer divisions. Some 700 were built and served from 1943. Its heavier partner in the Panzer force was the Hummel, fitted with the 15cm (5.91in) sFH 18.

In addition a small number of vehicles on a variety of chassis were built to carry the sIG 33 15cm gun. A further design (over 300 made) was the Sturmpanzer 4, sometimes known as the Brummbär, a well-armoured type carrying a 15cm StuH 43.

At the other extreme was the Karl Gerät, a 60cm (23.6in) siege mortar mounted on tracks. Six were made and they were provided with alternative 54cm (21.3in) barrels for longer range. The 60cm shell weighed 2,170kg (4,784lb) and could reach 6,580m (7,200yd).

STURMPANZER 4 BRUMMBÄR

The Brummbär ("Grizzly Bear") was an infantry support gun carried on a Panzer 4 chassis. Its 15cm howitzer fired a heavier (38kg/84lb) shell than the earlier StuG 3. It served successfully from 1943 and about 300 were built or converted.

WEIGHT: 28.6 tonnes
LENGTH: 5.93m (19ft 5in)
HEIGHT: 2.52m (8ft 3in)
WIDTH: 2.88m (9ft 5in)
ARMOUR: 100mm (3.9in) max.
ROAD SPEED: 40kph (25mph)
ARMAMENT: 1 x 15cm (5.91in) StuH 43; 38 rounds carried

Unarmoured Vehicles

Fighting armies consumed vast quantities of fuel and ammunition and fighting soldiers naturally needed to be fed. Transport vehicles to achieve these tasks were thus, if anything, more important than fighting equipment.

By the end of June 1944 the Allied forces had landed 150,000 vehicles in Normandy to support the 850,000 men by then deployed. Many of these were armoured fighting vehicles, but many more were soft-skin transport lorries, artillery tractors, repair trucks and other varieties. Another telling statistic is that during the invasion of the USSR in 1941 the Germans employed 2,000 different types of vehicle – and their army was only partly mechanized and still also had hundreds of thousands of draught animals. The importance of motor vehicles is thus obvious, as is the impossibility of detailing more than a sample of those used.

Probably the most famous transport vehicle of the war was the Jeep, originally designed by the Willys company and built mainly by Ford. To the US Army it was the "Truck ¼-ton, 4 x 4". Over 600,000 were made and many supplied to almost every Allied nation, in addition to their use with US forces. And as well as these there were vast numbers of ½-ton and ¾-ton vehicles from Ford, Dodge, Chevrolet and others in the light-truck class.

Other countries had equivalent equipment. The Soviets built GAZ-67 copies of the Jeep to supplement their Lend-Lease supplies. The Germans had the Kübelwagen, based on the original pre-war Volkswagen design. Even the Italians had a Fiat 508 type, which was used effectively in North Africa.

ARTILLERY TRACTORS

Jeeps were often used to tow anti-tank guns but all nations had specialized designs for the artillery-tractor role. British 25pdr guns were often towed by Morris C8 "Quad" tractors and heavier weapons by AEC Matadors. US heavy artillery units used several models of fully-tracked vehicle in the same role, which were surprisingly fast and had good cross-country capabilities. Many of these were made by the Allis-Chalmers company, previously known for its farm tractors.

Left: An American truck convoy on the Red Ball Express route in France in 1944.

Above: Tractors pull Soviet heavy artillery guns during the advance to Berlin in early 1945.

Above: A Scammel tank transporter leads a British column in the advance after El Alamein in 1942.

Soviet heavy artillery units also employed tracked vehicles produced by the tractor industry. German heavy towing vehicles included SdKfz 8 and 9 half tracks, which could also be used for tank-recovery duties.

Moving tanks to and from the battle area and recovering damaged ones in the field was a highly important support role. American tank-transporter types, also used by the British, included vehicles by Diamond T and Mack, while native British types included examples from Albion and Scammel.

CARGO VEHICLES

For simple cargo-carrying duties the USA's 2½-ton ("deuce and a half") design stands out. A truly massive 800,000 were built, mostly by General Motors, and they were supplied to all the Allies. By the end of the war the Soviets had more American trucks in use on the Eastern Front than indigenous ones.

Most countries had equivalent designs of roughly 3-tonne capacity, made by their own famous motor manufacturers, and usually a smaller number of larger vehicles up to the 10-tonne or 12-tonne class. Examples included Britain's 10-tonne Leyland Hippo or the Soviet YAG-10, an 8-tonne design.

The best-known use of transport lorries in the war was in the so-called Red Ball Express, set up by the Allied forces in France during 1944 in a desperate attempt to keep their advancing forces supplied. This used several thousand trucks running on a loop of one-way roads between St-Lô and Chartres. However, even this vast effort could not keep the armies going by mid-September 1944, when they were upward of 700km (435 miles) from the Normandy beaches, where their supplies were being landed.

TRANSPORT ANIMALS

The degree to which armies were mechanized varied considerably. Britain and the USA used motor transport whenever possible but in 1941, for example, the German Army had over 600,000 horses participating in Operation Barbarossa. In the course of the war the USSR probably lost 14 million horses, though many of these would not have been working with the Red Army. Even the British and Americans made significant use of transport animals, mainly mules, in particularly difficult terrain in Sicily, Italy and Burma. And over 1,000 elephants were used by the Allied forces in Burma.

Above: German transport crossing a river in Russia.

The Soviet Invasion of Germany

*As millions of refugees fled in terror before the Soviet advance, the Red Army ground
its way remorselessly from Warsaw to the Oder in the first months of 1945 and
from March began attacks to conquer Czechoslovakia and Austria.*

By January 1945 the Soviets were ready to resume their main attacks into Germany. Some 4 million men and masses of tanks, guns and aircraft were set to advance all along the front, from southern Poland to the Baltic coast of Lithuania.

Stalin had made it clear that he alone was in overall charge. Marshal Zhukov had been posted away from his position on the central staff to the front line, albeit to lead 1st Belorussian Front in the advance on Berlin; and Marshal Vasilevsky, the Chief of Staff, would also be given an operational command in February and replaced by a more junior officer.

The Soviet offensive began on 12 January with the strongest attacks coming from the bridgeheads west of the Vistula that

Above: Soviet gunners fighting on the outskirts of Breslau (Wroclaw). The city was surrounded in February 1945 but held out, on Hitler's orders, not surrendering until May.

had been held for some months by 1st Belorussian Front and Marshal Ivan Konev's 1st Ukrainian Front to the south. Within a week these troops had advanced well into Silesia, and Warsaw itself fell on the 17th.

The advance accelerated as the German front collapsed. For a time a panicked Hitler even put SS chief Heinrich Himmler, a man with no military training and less aptitude for generalship, in charge of Army Group Vistula, a new command that was meant to halt the rot. Instead, by early February, Zhukov's tanks had reached the Oder only 65km (40 miles) from Berlin.

TAKING EAST PRUSSIA

By then Germany's province of East Prussia had been largely overrun. The 2nd and 3rd Belorussian Fronts had attacked from the south and north-east at the same time as the Vistula offensive began. The southern attacks reached the Baltic near Elbing (Elblag) in early February, cutting East Prussia off from the rest of Germany. Almost all the rest of the province was taken by early April, when the capital Königsberg (Kaliningrad) surrendered.

Throughout this period there was a mass naval evacuation, with many casualties, of German troops and civilians from all

Left: Soviet engineers use a flame-thrower in their attack on Küstrin (Kostrzyn) on the Elbe in early 1945.

around the Bay of Danzig to ports like Kiel safely to the west. A few pockets on the coast remained in German hands until the surrender in May.

PAUSE ON THE ELBE

By February the Soviet spearheads seemed poised to drive on to Berlin, and probably could have done so relatively easily. However, for reasons that have never been clear, Stalin chose not to do this. Instead the Soviet forces spent several weeks taking control of Pomerania and southern Silesia. The best explanation seems to be that Stalin did not want the war to end before he had direct control of as much Polish and German territory as possible. And at this stage, with the Western Allies still fighting their way slowly to the Rhine, there seemed little prospect of them getting to Berlin first.

The Soviet forces south of Poland did little attacking in the first months of 1945 but did

Above: The Soviet advance was preceded by a tide of German refugees, rightly fearful of atrocities.

finish off the siege of Budapest in February. Bizarrely there now followed Germany's last significant offensive of the war. After the failure of the Battle of the Bulge, Hitler switched the elite Sixth SS Panzer Army to the Hungarian front and its attacks made limited gains in the Lake Balaton area in the first couple

of weeks of March. These were retaken immediately the Soviet offensives resumed on 16 March. In April the Soviets conquered much of Austria and by early May had moved well into Czechoslovakia.

The successful Anglo-American Rhine crossings in March had by then brought a new urgency to the operations on the main fronts. At the end of March Stalin finally gave orders for the decisive attack on Berlin.

Right: Members of the *Volkssturm*, or Home Guard, parade before Nazi chiefs in Posen (Poznań).

Heavy Mortars and Artillery Rockets

Germany's Nebelwerfer and the Soviet Katyusha were among the most-feared land weapons of WWII. Heavy mortars, too, generated awesome firepower concentrations and caused many casualties.

Heavy mortars and ground-to-ground rockets produced some of the most devastating sudden bombardments of the war. Rockets and mortars in fact had advantages over traditional artillery guns. They could bring down a heavy volume of fire quickly and fairly accurately on a target; and the more nearly vertical trajectory of their shells ensured that their fragmentation pattern was highly effective. More than half of the British casualties in north-west Europe in 1944–5 were inflicted by these weapons, for example.

In addition to the mortars used for infantry support, most armies also had heavier types, but the Germans and Soviets

Below: A Soviet 120mm (4.72in) mortar battery in action in the streets of Berlin at the end of the war.

made most use of these. The Soviet 120mm (4.72in) HM38 (based on a French Brandt design) fired a 16kg (35lb) bomb up to 6,000m (6,500yd). Other Soviet weapons were 160mm (6.3in) and 240mm (9.45in) designs. The German 12cm (4.72in) design was a copy of the Soviet HM38. The USA and Britain both used 4.2in (107mm) weapons. The American type was unusual in having a rifled barrel; most other mortars were smoothbore weapons.

ROCKET ARTILLERY

The Soviets and the Germans made more extensive use than other armies of rocket artillery. The German Nebelwerfer types included 6-tube 15cm (5.91in) and 5-tube 21cm (8.27in) designs as well as larger calibres. The Soviet Katyushas were

often truck mounted and included the M8 firing 32 x 82mm (3.23in) rockets and the M13 firing 16 x 132mm (5.2in) rockets. These latter had an 18kg (40lb) warhead and could reach a range of 8,500m (9,300yd).

Rocket weapons often had a distinctive noise when fired: Anglo-American troops knew the Nebelwerfer types as "Moaning Minnies"; Red Army troops called their Katyushas "Stalin's Organs".

SdKfz 4/1 MAULTIER

The SdKfz 4 was a half-track version of Germany's standard 3-tonne trucks. It was designed to cope with difficult ground on the Eastern Front. About 300 were built to carry Panzerwerfer 42 rocket launchers.

WEIGHT: 7.1 tonnes
LENGTH: 6m (19ft 8in)
HEIGHT: 2.5m (8ft 2in)
ARMOUR: 8mm (0.3in)
ROAD SPEED: 40kph (25mph)
ARMAMENT: 10 x 15cm (5.91in) Panzerwerfer 42 rockets

V-Weapons

Hitler thought that these "Retaliation Weapons" would win the war for Germany, but despite their sinister reputation their effectiveness was more limited. Producing them also cost the lives of thousands of slave labourers.

Cruise missiles and ballistic missiles are part of the everyday military vocabulary of the 21st century; their ultimate ancestors were the German Fi-103 and A-4, which saw much use in WWII. Both types had a variety of codenames and designations: most are commonly known by the V designation (V for *Vergeltungswaffe*, or "retaliation weapon"), first coined by German propaganda and then officially endorsed by Hitler.

FIESELER 103/V-1

FUSELAGE LENGTH: 6.65m (21ft 10in)
WINGSPAN: 5.33m (17ft 6in)
ENGINE: Argus 109-014 pulse jet; 310kg (680lb) max. thrust
MAX. SPEED: 670kph (415mph)
MAX. RANGE: 200km (125 miles)
WARHEAD: 850kg (1,870lb)

V-1 FLYING BOMB

The Fieseler 103, or V-1 flying bomb, was a small pilotless aircraft powered by a pulse-jet engine, fitted with an auto-pilot to guide it to its target. Tested from 1942, it was put into action in June 1944, a few days after the D-Day landings.

In the early stages most of the missiles fired were targeted on London. Of the roughly 8,600 launched before the French bases were overrun by the Allied armies, about a quarter reached their targets. Later in the war Antwerp and other Belgian cities were targeted, again with modest accuracy; about half of the missiles launched landed within a dozen kilometres of their targets.

Although the V-1 was cheap to make (around 5,000 marks each, 2.5 per cent of the price of a V-2), the 30,000 made took up more than half of German explosives production in 1944–5 and inflicted fewer than 7,000 fatalities on the British people.

V-2 ROCKET

The A-4/V-2 ballistic missile was an altogether more high-tech weapon. Its liquid-fuel rocket carried it 80km (50 miles) into the stratosphere before it fell at supersonic speed onto its target. Unlike the V-1 it could not be intercepted by fighter aircraft or anti-aircraft gunfire.

About 3,500 were fired at London and other cities from September 1944, carrying in all

less explosive power than a single large Allied bombing raid on Germany in the same period. More slave labourers died in the Nazi factories making the V-2 than were killed by the missile attacks. With a more powerful warhead, the story might have been different. However, in this vein, it has also been calculated that the development cost of the V-2 was about the same as the American expenditure in making the atom bomb.

A-4C/V-2

LENGTH: 14m (45ft 11in)
FUSELAGE DIAMETER: 1.64m (5ft 5in)
ENGINE: Liquid-fuel rocket
FLIGHT TIME: 330 seconds
MAX. RANGE: 314km (195 miles)
WARHEAD: 730–975kg (1,600–2,150lb)

Battle of the Bulge

Hitler aimed to repeat the triumph of 1940 in an attack through the Ardennes region to cut the Allied armies in two. Instead Germany's last reserves were defeated in a series of desperate winter battles.

By the late autumn of 1944 the German Army had recovered some of its strength after the disasters of both the summer in Normandy and the Eastern Front. However, the American armies were pushing forward slowly in eastern France and Belgium, while the British had finally succeeded in clearing the Scheldt estuary, so that the great port facilities of Antwerp could at long last begin to alleviate the Allied supply problems.

Hitler decided to use the assembling German reserve force in the west, where he was convinced a major success would bring quarrels between

Below: Draped with machine-gun ammunition and much other equipment, an SS trooper prepares to attack during the Ardennes battle.

Above: An M36 tank destroyer in the Ardennes. The M36's 90mm (3.54in) gun was the best US anti-tank weapon in 1944–5.

Britain and the USA, persuade them to make peace and perhaps even join him in fighting the Soviets. The plan was to advance through the Ardennes region, just as in May 1940, and then drive in a north-easterly direction to Antwerp, cutting the Allied front in two. All this was fantasy but Hitler insisted the attack would go ahead, despite protests from the commanders who would actually have to carry it out.

Allied intelligence was aware of the assembly of German reserves but believed they were intended as a counter-attack force against any future Allied advances. The Ardennes region was indeed one of the weakest parts of the Allied line, held by an unfortunate mix of resting veterans and inexperienced newcomers. The Allied commitment to advances to the north and south of the region also meant that there were few

reserves available to use elsewhere. Altogether the Germans assembled ten Panzer divisions, equipped with many of the newest heavy tanks, and a number of newly raised *Volksgrenadier* infantry units. Three armies were deployed; together they formed Field Marshal Walther Model's Army Group B.

INITIAL GERMAN GAINS

The Germans attacked on 16 December and achieved complete surprise. They quickly broke through the Allied line all along the attack front, while small groups of special forces penetrated deeper into Allied territory, spreading confusion and panic. A few Allied reinforcements were quickly sent to the area and they, and the survivors of the original front-line force, established themselves especially around the towns of St-Vith and Bastogne. They were both important road junctions, particularly vital for movement in an area of steep and densely wooded hillsides.

By the 19th the top Allied commanders were taking the situation in hand. All Allied attacks elsewhere were halted and new arrangements to deal with the threat established. Field Marshal Montgomery was put in charge of the Anglo-American forces north of the German advance, and General Omar Bradley of the US forces to the south. Part of Patton's US Third Army changed front with

Above: Men of US 75th Division advancing toward St-Vith during the Allied counter-attack in January 1945.

BODENPLATTE

The Luftwaffe planned a knock-out blow – Operation Bodenplatte – against Allied air bases in Belgium to accompany the Ardennes offensive. It was finally delivered on 1 January 1945 with something approaching 900 German aircraft taking part, many flown by novice pilots. They surprised a number of Allied airfields and destroyed about 150 aircraft, mostly on the ground, but about 300 of the German planes and many of the most experienced pilots were shot down. It was yet another serious defeat for the Luftwaffe.

BATTLE OF THE BULGE
The German plan for the Ardennes offensive was to split the Allied armies apart and capture their principal supply port, Antwerp.

astonishing speed and began attacking north to relieve Bastogne and reduce the bulge the Germans had now driven into the Allied line.

Up to this point bad weather had kept Allied air support to a minimum, but on the 22nd it cleared. The German supply system was already stretched; now both it and the front-line forces came under continuous attack. A few German troops almost reached the River Meuse by late on the 24th and would go no further. Their generals were already begging Hitler's permission to call off the attack (and being refused).

THE ALLIED RESPONSE

After an epic defence, Bastogne was relieved by Patton's advance on 26 December. Major attacks from the north of the Bulge began on 3 January, and it had largely been recovered by the middle of the month. The Germans lost about 100,000 men and most of the tanks used in the operation. Allied losses were similar in number but theirs could be replaced. Germany's could not.

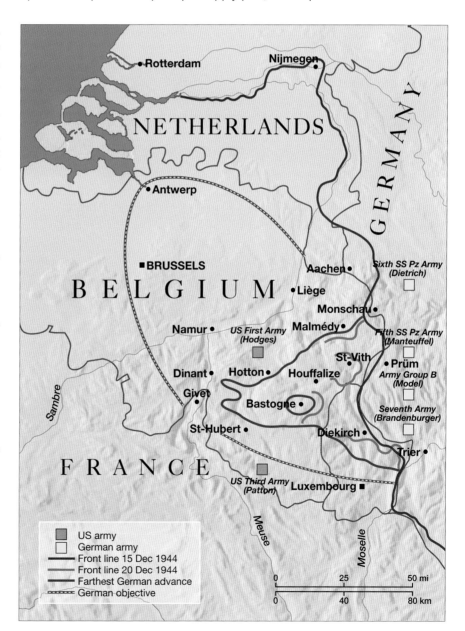

Jet Aircraft

Although the piston-engined aircraft in service in 1944–5 were vastly superior to those used at the start of the war, they were clearly outclassed by the new turbojet types, even though all of these had reliability issues and other problems.

Although the power of aircraft piston engines was improved very substantially during the course of WWII, it was also clear that there was more potential for the future in rocket and turbojet propulsion. All the major combatants were working on such designs by 1945, but only Germany and Britain used jet aircraft in combat before the end of the war; the USA had jets in service but not yet deployed on operations.

FIRST JETS

The first workable jet engines were made in 1937 in separate research programmes in Britain and Germany. At that stage the British design by Frank Whittle was more advanced, but Heinz von Ohain's engine was the first to fly, in the Heinkel (He) 178 in 1939. Whittle's Gloster E.28/39 first flew in 1941. Neither of these was intended as a combat aircraft, but two

GLOSTER METEOR F 1

The Gloster Meteor was the first operational Allied jet. The illustrated example, an F 1, is in flight over Kent in August 1944, days after the type's combat debut. The later F 3 had better Derwent engines and improved aerodynamics.

SPEED: 670kph (417mph)
RANGE: 880km (550 miles)
CREW: 1
ENGINES: 2 x Rolls-Royce W2B Welland turbojets
ARMAMENT: 4 x 20mm (0.79in) cannon

designs which were took to the air in 1941–2: the Messerschmitt (Me) 163 and Me 262.

The Me 163 came from a completely different line of development. With an entirely unconventional arrowhead shape and powered by a Walther rocket engine, it could reach an incredible 960kph (596mph). Disadvantages were that it only carried fuel for about ten minutes of flight; that the fuel itself was dangerously prone to explosion; and that after its brief flight the aircraft had to glide back to base and land using skids, not a proper undercarriage. It began operations in mid-1944 and achieved a few successes against Allied bombers, but accidents and other problems were common.

The turbojet and Me 262 offered better prospects, though it took two years from the type's first flight in July 1942 for it to reach combat. This was mainly caused by engine and other faults, but Hitler's order that it was to be adapted as a fast bomber certainly did not help.

Although perhaps 1,400 were built, far fewer reached the squadrons and they suffered fuel shortages and other problems. They could outfly any Allied aircraft and were well-armed for bomber interception, but many were shot down while

Left: The Me 163 rocket fighter was stunningly fast but probably shot down fewer than 20 Allied aircraft.

landing or taking off from their bases when they were slow-moving and vulnerable.

Two more German jets were built and made operational but saw little combat. Over 200 Arado 234 bombers were produced in 1944–5 but saw only scattered action, notably against the Rhine Bridge at Remagen.

Some 50 of a simplified fighter type, the He 162 Volks-jäger (or "People's Fighter") were built in the final months of the war. This design was meant to be suitable for operation by inexperienced pilots, but by the time it was available there was little fuel to be found. It probably flew a handful of combat missions in April 1945.

ALLIED TYPES

On the Allied side the American Bell P-59 Airacomet first flew in October 1942 but met repeated problems and was never taken into operational service. Two more US designs, the Lockheed P-80 and the Ryan FR-1 (this

type having both a piston and a jet engine fitted), were being deployed when Japan surrendered but did not see combat.

Britain's Gloster Meteor flew in 1943 and came into operational use in July 1944. Initially F 1 models were employed in the home-defence role, operating against V-1 flying bombs, but a few much improved F 3s were deployed to bases on the Continent in 1945. Like the other early jets, it suffered from many teething problems and successive production batches showed numerous minor improvements to address these.

ARADO 234

The Arado 234 first flew in June 1943. Early production versions, designated Ar 234B (see specification below), had two Junkers engines. The illustrated example is a 234C, with four BMW 003A engines, which was significantly faster but was still in prototype form when the war ended.

SPEED: 742kph (461mph)
RANGE: 1,100km (680 miles) with bomb load
CREW: 1
ENGINES: 2 x Junkers Jumo 004B-1 turbojets
ARMAMENT: 1,500kg (3,300lb) bombs; usually no guns

Right: A bomb-armed Me 262. Hitler insisted that the type should be developed as a fast bomber.

Anti-tank Guns, 1942–5

Higher muzzle velocities and new types of ammunition ensured that anti-tank guns still posed a formidable threat to even the monster heavy tanks produced for the final battles of the war.

The increase in power seen in anti-tank guns of the early-war years continued during the second half of the war, though with the improvements being derived more from new types of ammunition rather than large increases in calibre. Armies also found that towed anti-tank guns in the larger calibres could be clumsy in action and more difficult to conceal;

NASHORN

The Nashorn ("Rhinoceros"; also known as the Hornisse, or "Hornet") was a very effective tank destroyer armed with the long-barrel 8.8cm gun. The basic vehicle was a modified Panzer 4 chassis. It was first used at Kursk in 1943.

WEIGHT: 24 tonnes
LENGTH: 7.17m (23ft 6in)
HEIGHT: 2.65m (8ft 8in)
WIDTH: 2.8m (9ft 2in)
ARMOUR: 30mm (1.2in) max.
ROAD SPEED: 42kph (26mph)
ARMAMENT: 1 x 8.8cm (3.46in) PaK 43; 25 rounds carried

accordingly they turned increasingly to self-propelled (SP) weapons in the anti-tank role.

NEW AMMUNITION

In ammunition, simple solid-shot armour-piercing (AP) designs were found to be prone to shattering on impact, because of the higher velocities being used. They were also more likely to glance off the sloped armour that more and more tanks included. This led to the introduction of capped rounds (APC) to achieve better impacts and ballistic caps on top of these (APCBC) to restore ideal shapes.

Composite rounds with a dense penetrating core encased in a lighter carrier (armour-piercing composite rigid – APCR; or high velocity armour-piercing in the US designation – HVAP) were also introduced, as well as a better version of these in which the light carrier fell away on leaving the barrel (armour-piercing discarding sabot – APDS). These had an ideal combination of lightness to accelerate quickly in the gun barrel but with a hard and dense munition with good carrying power, penetration and accuracy. Various guns had a different range of ammunition developed at different times during the

Below: A 17pdr (76mm) anti-tank gun of the 2nd New Zealand Division in action at Monte Cassino in Italy in early 1944.

war; APDS was a British speciality introduced for the 6pdr (57mm) in the spring of 1944, for example.

Earlier weapons like the 6pdr remained in use to the end of the war, in part thanks to improved ammunition, but also because they could still be effective against most opponents at shorter ranges. In British service, however, the 6pdr was supplemented from 1942 by the 17pdr (76mm). The essentially identical US 57mm was supplemented by the 3in (76mm) M5 and the dual-purpose version of the 90mm (3.54in), originally made as an anti-aircraft gun.

SELF-PROPELLED GUNS

All of these weapons featured on SP mounts. Indeed the towed and SP guns served together in the American tank-destroyer units. The most notable of these were the M10 (with the 3in gun) and the M36 (90mm), as well as a lighter, faster design, the M18 Hellcat, which mounted the same 76mm weapon as later US Sherman tanks. For Britain, the 17pdr gun was mounted in an M10 variant – the Achilles – and in a vehicle derived from the Valentine tank – the Archer.

Soviet anti-tank weapons included the ZiS-3 76.2mm (3in) field gun, which had a respectable AP performance. This gun was made by the tens of thousands, along with the very formidable 100mm (3.94in) BS-3 introduced in 1944. Most

Soviet SP guns with a significant anti-tank capability were more heavily armoured SU-series assault guns.

As well as its assault guns, Germany also had various *Panzerjäger* (or "tank-hunter") types, as well as towed weapons of course. The 7.5cm (2.95in) PaK 40 towed gun was introduced in late 1941 and served to the end of the war. There was also an improved version of the "Eighty-eight" – the 71-calibre 8.8cm (3.46in) PaK 43.

The *Panzerjäger* vehicles included several types known as the Marder. These were based on a variety of chassis and armed with either the German PaK 40 gun or captured Soviet 76.2mm weapons. They were introduced in 1942–3. The later Hetzer was the most numerous *Panzerjäger*, with some 2,500 produced, and also carried a 7.5cm gun. The Nashorn was more formidable, carrying the long-barrelled PaK 43, but only some 500 were made.

M36 GMC

The M36 Gun Motor Carriage entered service in 1944. Several variants existed, built with different engines and hulls. All carried the anti-tank version of the 90mm anti-aircraft gun and could defeat Panther or Tiger tanks.

WEIGHT: 28.1 tonnes
LENGTH: 6.15m (20ft 2in)
HEIGHT: 2.72m (8ft 11in)
WIDTH: 3.04m (10ft)
ARMOUR: 50mm (2in) max.
ROAD SPEED: 48kph (30mph)
ARMAMENT: 1 x 90mm (3.54in) M3; 47 rounds carried

Victory in the West and Italy

Allied commanders anticipated a hard struggle to conquer Germany. Instead, after they crossed the Rhine, German resistance collapsed within weeks. However, the great prize of Berlin was left for the Soviets to take.

After the disappointments of the autumn of 1944 and the hard fighting of the Battle of the Bulge at the turn of the year, the Allied commanders did not expect to finish the campaign against Germany easily. In particular they thought that crossing the Rhine against tough German opposition would be difficult, but events proved them wrong.

British and Canadian troops opened the Allied offensive at the north end of the line in early February, fighting a vicious battle through the Reichswald forest to close up to the Rhine. US Ninth Army on their southern flank was meant to join the attack shortly after but was delayed by extensive flooding, deliberately caused by the Germans. By early March, however, Ninth Army was on the

Above: Men of British Eighth Army near Ferrara during the final Allied attacks in Italy in April 1945.

move and had been joined by US First and Third Armies farther to the south still.

Within days the Allied troops had reached the Rhine everywhere north of Cologne, and on

7 March US First Army reached Remagen some 45km (28 miles) to the south.

Better still, the Americans captured the railway bridge over the river there before it could be destroyed by the retreating Germans, and immediately started rushing troops across. Finally, the last Allied armies, US Seventh and French First, overran almost all the remaining German-held territory west of the river by mid-month.

Montgomery's British forces made their long-planned assault crossing of the Rhine from late on the 23rd around Wesel, with much less difficulty than anticipated. Originally this had been conceived as the main Allied advance but the crossing at Remagen had been joined on the 22nd by another – by Third Army south of Mainz. Patton was determined to steal Monty's glory. A few days earlier General Eisenhower had already decided to modify Allied plans by strengthening these attacks south of the Ruhr.

ENCIRCLING THE RUHR

Hitler was still issuing his usual orders that there should be no retreat and fanatical resistance. However, the reality was that increasing numbers of German troops were now only too glad to surrender to the Western Allies

Left: American troops at the bridge at Remagen, the first crossing point over the Rhine captured by the Allies.

in order to keep themselves safe from the Russians. New Allied advances at the end of March encircled the whole Ruhr area and the troops there – over 300,000 strong – surrendered in mid-April.

By this stage Soviet troops were on the Oder, only 65km (40 miles) from Berlin and seemed well-placed to capture it. In mid-March Eisenhower therefore decided that his forces would not try to reach Berlin but would concentrate their advances farther south against a feared (though actually non-existent) National Redoubt that fanatical Nazis were supposedly building for a last stand in southern Germany.

In fact the Anglo-American armies were now advancing at speed all along the front, while the Soviets were not yet ready for their final drive to Berlin. US troops reached the Elbe west of the German capital in mid-April and halted there, with first contact with the Soviets coming a little to the south-west on the 25th at Torgau.

LAST BATTLES IN ITALY

After the capture of Rome in June 1944, Allied troops had been taken from Italy for the invasion of southern France. The remaining Allied units continued a slow advance into early 1945. In April they renewed their attacks, now with more success. German forces in Italy surrendered on 2 May, and on 4 May the advancing Allied troops linked up at the Brenner Pass with US Seventh Army coming down through Bavaria.

By early May most of what would become West Germany was in Anglo-American hands

and the formal surrenders began here, too. German forces in the north capitulated to Montgomery on the 4th, followed by a complete German surrender that was signed at Eisenhower's headquarters on the 7th.

Above: Tanks and troops of US Third Army crossing the Rhine near Koblenz on a pontoon bridge.

Below: An Allied M3 half track on the advance through the ruins of a German town in early 1945.

Infantry Anti-tank Weapons

By 1945 infantrymen in the best-equipped armies had man-portable weapons that could destroy even the heaviest tank – if they were brave enough to get desperately close to the metal monsters.

Specially powerful rifles had been used as anti-tank weapons almost from the moment tanks were introduced in WWI. Versions of these were still in service when WWII broke out but were soon made obsolete by improvements in tank armour.

From 1942–3, Britain, the USA and Germany introduced weapons using the hollow-charge principle to focus explosive power. These gave the infantry weapons that could kill a tank, or smash open a pillbox or other fortification.

ANTI-TANK RIFLES
In the early-war years anti-tank rifles in service included the British Boys type, in 0.55in (14mm) calibre, and two Soviet designs – the PTRD 1941 and PTRS 1941 – both in 14.55mm (0.57in) calibre. These all had

Above: A young German soldier waits in a ruined house with his Panzerfaust at the ready.

similar performance, being capable of penetrating some 25mm (1in) of armour at 400m (440yd). Smaller-calibre types included the Polish Maroszek Wz 35 (also used in significant

numbers by the Italians) and the German Panzerbüsche 38 and 39, all in 7.92mm (0.312in).

ROCKET LAUNCHERS
The US Army had no anti-tank rifle when it entered the war but soon acquired something better – the 2.36in (60mm) M1 Rocket Launcher (called the Bazooka). This could destroy most contemporary Axis tanks and had a range of up to 400m (440yd), though accuracy at this distance was far from certain. Later versions had the calibre increased to 3in (76.2mm) with improved armour-piercing performance.

Germany's 8.8cm (3.46in) Panzerschreck (or "Tank Terror") was essentially a copy of the Bazooka. Britain's Projector Infantry Anti-Tank (PIAT) looked clumsy and primitive, but it was able to fire a useful 1.4kg (3.1lb) bomb.

Germany also used the even more innovative Panzerfaust (or "Tank Fist"), a light and simple single-shot weapon with the firing tube designed to be discarded after use. The longest-range Panzerfaust variant could only reach 100m (110yd), but its power and ease of use made it a formidable threat to any tank, especially in close country or a built-up area.

No other army developed such weapons during WWII.

Left: Later in the war the Soviets often used anti-tank rifles, as here, against enemy positions, not tanks.

Mines and Other Defences

Mines, and their close relatives booby traps, are sometimes called silent soldiers,
lying patiently in wait for an enemy. They were a threat on every battlefield of the
war, capable of killing in battle and long after fighting had moved elsewhere.

There were two main categories of mine that were in use during WWII – anti-personnel and anti-tank – the former usually being smaller and requiring a lighter force to set them off, the latter larger and heavier. Minefields could be set in any open area and contain one, or more often both, types; and individual or small numbers of mines could be used for point defence or as booby traps virtually anywhere.

Troops hated enemy mines, with some justification. Anti-personnel mines were often designed more to maim than to kill, since a wounded soldier often got assistance, taking perhaps several combatants out of the firing line.

The design of mines also often inspired fear. Both Germany and the USA used mines that had two charges – one to throw the mine into the air when it was disturbed; and a second main charge to detonate it there. Germany's Schrapnellmine, for example, blasted out some 350 shrapnel balls from a position roughly 1.5m (5ft) above ground, with devastating effect.

Anti-tank mines could be both bar- and discus-shaped. The Italian forces were the first to use them, in the Western Desert in 1940. Common designs included the Soviet TM/39 which carried a 3kg (6.6lb) charge or the slightly heavier British Mark V.

Germany and the USSR were the most prolific users of mines. As well as the most often encountered Tellermine anti-tank designs, Germany also had a variety of anti-tank and anti-personnel types in glass, plastic and wooden cases, all designed to make detection harder. Many mines of all nations were also fitted with anti-lifting devices to make their removal more difficult and dangerous.

Right: Soviet engineers laying mines to block a German advance on the Eastern Front in 1942.

Above: A Canadian engineer preparing to clear an enemy anti-tank mine in Italy.

The Fall of Berlin

In the last weeks of the war the ruin that Hitler had brought to Europe was visited in yet a more terrible form on Berlin, as the Soviet armies rampaged through the city above his bunker and across much of Germany.

From the start, WWII had been a total war, prosecuted with the utmost violence and cruelty, and its final great battle saw this trend continue to the last. At the end of March 1945 General Eisenhower had told the Soviets that his forces would stop short of Berlin and make no attempt to capture the city. Stalin did not believe him and hurried forward the Red Army's preparations to get there first.

ODER BREAKTHROUGH

The Soviet attack began on 16 April with some 2.5 million men and a truly astonishing 16,000 artillery weapons deployed. The Germans had perhaps one million men, but many of them were over-age or very young. They had nothing like the scale of equipment of their enemies, and even when they did have tanks they had scarcely any fuel

---

KEY FACTS

PLACE: Berlin

DATE: 16 April–2 May 1945

OUTCOME: Soviet troops advanced from the River Oder to conquer Berlin; Hitler committed suicide.

---

for them. Even so, the initial Soviet attacks, by Zhukov's 1st Belorussian Front, against the Seelow Heights west of Küstrin failed to break through, in part because Army Group Vistula was now led by an acknowledged defensive expert, General Gotthard Heinrici.

However, by 20 April the German resistance had inevitably fallen apart. Zhukov's surge toward Berlin and round its

north side had also been joined by a major drive by Konev's 2nd Belorussian Front over the Neisse a little to the south.

To ensure the fastest progress Stalin gave both his marshals permission to take Berlin, according to which got there first. Zhukov's troops reached the edges of the city on the 21st and the two fronts met to complete the encirclement west of Potsdam on the 25th. On the same day Konev's men made contact with US troops at Torgau on the Elbe to the south.

Over the next few days the Soviets fought their way from street to street and house to house. On 30 April they stormed the Reichstag building, only 400m (440yd) from Hitler's command bunker. On the 29th Hitler had named Admiral Dönitz as his successor and then killed himself. The fighting went on a little longer, but on 2 May the Soviet fronts linked up from north and south and the last of the garrison surrendered.

In the meantime the remaining troops of the German Ninth Army, part of the defence force on the Oder a couple of weeks before, were attacking desperately to the west to escape the Soviets and surrender to the Americans. They knew only too well the fate that probably would await them in Soviet captivity. Since the Soviets had

Left: Raising the Red Flag over Germany's parliament, the Reichstag.

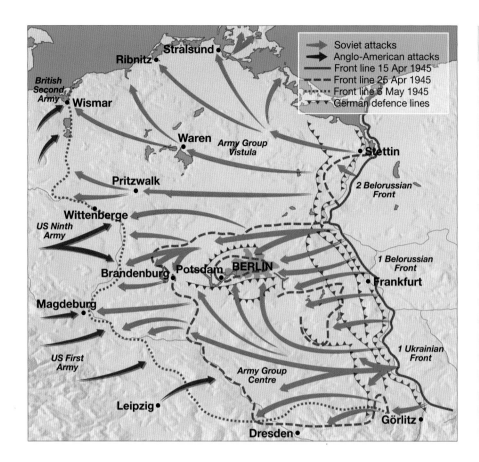

Konev (1897–1973) was a tough and competent Soviet commander who held senior posts throughout the war. He played a notable part as a front commander in the defence of Moscow in late 1941 but then, and later in his career, he was somewhat overshadowed by Zhukov, with whom he had a great rivalry. Konev's public reputation grew with his part in the Ukraine and Polish battles of 1943–4. In 1945 he shared in the conquest of Berlin, inevitably with Zhukov, but Konev finished the war off on his own by leading the capture of Prague.

BATTLE OF BERLIN
The Soviet advance from the Oder took only about three weeks to overwhelm Berlin.

entered Germany, their advance had been marked by a rampage of murder, rape and drunken looting, which now reached a terrible climax in and around the enemy capital.

GERMANY SURRENDERS
With Hitler dead, the end was certain. German troops in Italy surrendered on 2 May, in north Germany on 4 May and in south Germany on 5 May. On 7 May Dönitz's representatives signed the overall surrender at Eisenhower's HQ at Reims. Stalin suspiciously insisted that the signing be repeated late on the

8th in Berlin. The Western Allies, however, celebrated VE-Day (Victory in Europe) on the 8th, the Soviets on the 9th.

Even then the killing was not over. Some of the last German units holding out by late April were in Czechoslovakia. On 5

May the people of Prague rose against the Germans, who fought back fiercely. Over the next week Soviet units closed in from the east and north, and the final German force surrendered on the 13th. The war in Europe was finally over.

Right: Soviet Josef Stalin tanks in front of the captured Reichstag.

The Battle of Leyte Gulf
Japanese kamikaze pilots at a
pre-mission ceremony.

**Landing Craft and Amphibious
Vehicles** A wrecked Japanese
landing barge on Saipan.

Submarines and Bombers
Admiral Lockwood aboard the
submarine USS *Balao*.

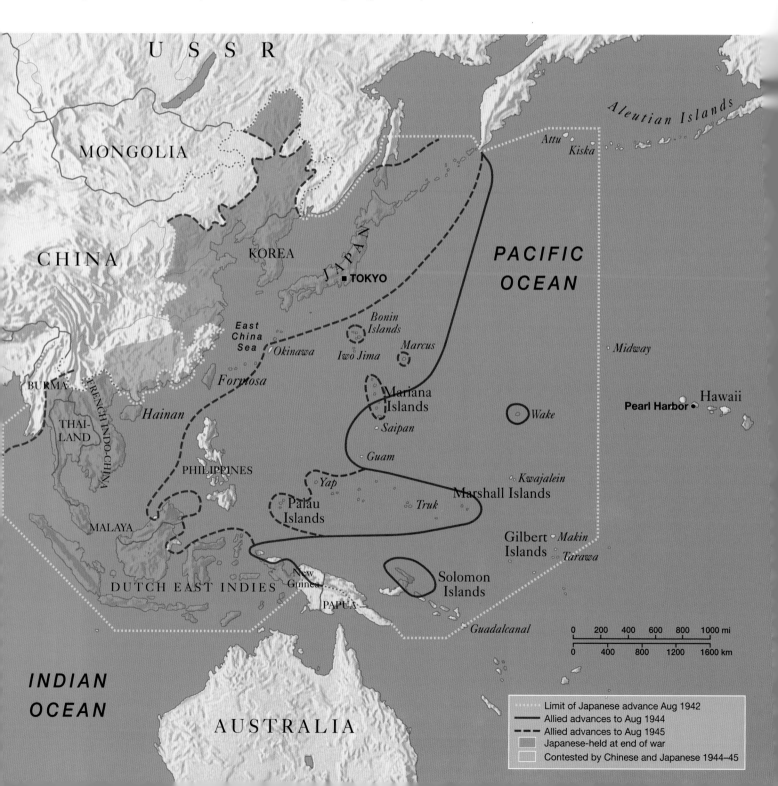

U S S R

MONGOLIA

CHINA

KOREA

JAPAN

■ TOKYO

Aleutian Islands

Attu

Kiska

PACIFIC
OCEAN

East
China
Sea

Okinawa

*Bonin
Islands*

Iwo Jima

Marcus

Midway

Formosa

BURMA

FRENCH INDO-CHINA

Hainan

THAI-
LAND

*Mariana
Islands*

Saipan

Wake

Pearl Harbor ○ Hawaii

PHILIPPINES

Guam

Kwajalein

Yap

Truk

Marshall Islands

MALAYA

*Palau
Islands*

DUTCH EAST INDIES

New
Guinea

PAPUA

*Solomon
Islands*

Gilbert *Makin*
Islands
Tarawa

Guadalcanal

INDIAN
OCEAN

AUSTRALIA

| 0 | 200 | 400 | 600 | 800 | 1000 mi |
| 0 | 400 | 800 | 1200 | 1600 km |

Limit of Japanese advance Aug 1942
Allied advances to Aug 1944
Allied advances to Aug 1945
Japanese-held at end of war
Contested by Chinese and Japanese 1944–45

DESTRUCTION OF THE JAPANESE EMPIRE

By 1943 the vast expansion of America's military strength was taking effect. Despite large increases in its own war production Japan could not begin to compete. From relatively small beginnings in New Guinea and the Solomons, the Allied counter-offensive soon developed in pace and strength. The US Navy and Marines began their island-hopping advance across the Central Pacific and MacArthur's Allied South-west Pacific Command surged toward the Philippines. By mid-1944 a relentless submarine campaign was making Japan's vaunted empire practically useless to the homeland, which was itself coming under heavy bombing attack. On the Asian mainland Japan could still advance against weak Chinese forces, but plans to invade India from Burma had been smashed by the British–Indian Fourteenth Army.

The second half of 1944 and early 1945 saw the ring around Japan tighten with the fall of the Philippines, Iwo Jima and Okinawa. The desperate expedient of suicidal kamikaze attacks did little to slow the American advance. Allied plans for an invasion of Japan were well in hand by August 1945, when everything changed with the atom bomb attacks on Hiroshima and Nagasaki and the crushing Soviet offensive on mainland Asia. Japan's surrender followed within days.

Naval Bombers and Torpedo Aircraft A US Navy Curtiss SB2C Helldiver in flight.

Conquest of the Philippines US Navy landing ships unloading troops and supplies on Leyte.

The Japanese Surrender The mushroom cloud of an atomic explosion rises above Nagasaki.

New Guinea and the Solomons

Japan's troops began paying the price for their reckless early-war conquests.
Garrisons were isolated and then either picked off or left to starve as the Allied
forces pushed past in their counter-offensive toward the Philippines.

After the triumph at Midway, American strategists knew that they could move to the attack in the Pacific, but there was fierce debate on how to develop the counter-offensive. General MacArthur, now established in Australia at the head of the Allied South-west Pacific Command, argued forcibly for an advance through the Solomons and along the north coast of New Guinea toward eventual fulfilment of his promise to return to the Philippines.

Admiral Chester Nimitz, commanding the Pacific Ocean Areas from his base in Hawaii, wanted to advance via the island groups of the Central Pacific. Like MacArthur's, this was a strategy that favoured his own command and his own service, the US Navy, but in truth it was also likely to be more direct and economical in casualties and material. However, American resources were becoming so ample that both plans could be

Above: Stuart tanks leading a search for Japanese positions, New Georgia Island, August 1943.

pursued. Indeed it would have been very difficult politically for the President and his advisers, with whom the decision ultimately rested, to have chosen one strategy exclusively.

AIR SUPERIORITY

All the Allied advances were based on air power, an area in which they were clearly superior by the end of 1942. This was

well demonstrated in the Battle of the Bismarck Sea in the first days of March 1943. A Japanese convoy had sailed from Rabaul to reinforce their forward positions on New Guinea around Lae but was attacked en route. A dozen ships were sunk and around 4,000 soldiers drowned.

The attacks were made possible by code-breaking information, which would play a vital part throughout the remainder of the campaign. Time and again the Allied forces would attack weakly defended locations and quickly build air bases to ensure dominance over any stronger Japanese forces nearby.

Major Japanese positions were neutralized by air attacks and then in effect ignored, left to "wither on the vine" while the Allied advance moved on elsewhere. This technique was even applied to the principal Japanese strongholds of Rabaul and Kavieng, which eventually were left isolated and impotent (though early plans had provided for their capture).

NEW GUINEA

In the first phase of the advance the Japanese were drawn forward, by air landings and naval attacks by Australian and American forces, toward Wau and Salamaua – and then kept off balance by other forces leap-

Left: B-25 bombers of US Fifth Air Force attacking a Japanese airfield on New Guinea, February 1943.

frogging on to the Markham Valley and Huon Peninsula areas. In the second phase, in April–May 1944, landings at Aitape and Hollandia isolated tens of thousands of Japanese troops around Wewak. The final landings on western New Guinea and nearby islands established bases to support the Marianas and Philippines operations, which were to follow.

THE SOLOMONS

Again the same techniques were followed in the Solomons campaign. This was most clearly seen in the landings on Bougainville, which began on 1 November 1943. The island had a garrison of some 60,000 Japanese troops concentrated in the south around Buin. Instead of attacking directly, US Marines landed 120km (75 miles) away on the east coast at Empress Augusta Bay. By the time the Japanese had crossed the jungle to the Allied beachhead it was easily strong enough to repel their attacks.

Although fighting on Bougainville continued to the end of the war, by then being conducted on the Allied side mainly by Australian forces, the

starving Japanese garrison, like others in the region, was contributing nothing of value to Japan's war effort.

Above: US Army troops advance cautiously through dense jungle on Vella Lavella Island in the New Georgia group, September 1943.

Above: Australian troops in a landing craft shortly before the attack on Lae, September 1943.

GENERAL DOUGLAS MACARTHUR

MacArthur (1880–1964) was the top US Army general in the Pacific throughout the war. He commanded the American and local forces in the Philippines in 1941 but was quickly defeated by the first Japanese attacks. MacArthur was ordered to leave and take over a new Allied South-west Pacific Command, which he led successfully in the New Guinea campaign until the return to the Philippines in 1944–5. Like many top commanders, he was a great egotist and self-publicist, but he was nonetheless also a talented and effective general.

Below: General MacArthur in the Philippines in early 1945 after his promotion to General of the Army.

Seaplanes and Naval Support Aircraft

Distant patrols far out across stormy oceans, perhaps ending in a sudden attack on an enemy submarine, were the stock-in-trade of maritime aircraft. As in many other military duties there might be hours of boredom, then moments of desperate fear.

Maritime patrol aircraft included large long-range seaplanes, smaller floatplanes (often launched from a catapult aboard a ship other than an aircraft carrier) and, especially later on in the war as radar equipment was developed, land-based types used for long-range anti-submarine work. The leading maritime powers (the UK, USA and Japan) made most use of such aircraft, but all nations with a navy of any size employed them to some extent.

FLOATPLANES

There were numerous designs of this type in service. France, the USA and Germany each had more than ten types, so only a few can be highlighted here.

The most numerous American design was the Vought OS2U Kingfisher (over 1,500 built). It had a range of some 1,300km (800 miles) and flew from catapults aboard many US Navy battleships and cruisers.

The two notable Japanese machines were the biplane Mitsubishi F1M "Pete", which had a surprisingly good performance, and the more capable Aichi E13A "Jake" with a 2,100km (1,300 miles) range.

Germany's best aircraft in this class was the Heinkel 115, but the Arado 196 also saw significant use.

Below: Vought OS2U Kingfisher floatplanes on their catapults aboard the heavy cruiser USS *Quincy*, 1944.

Britain's most common small floatplane was the Supermarine Walrus, an obsolescent pusher biplane, which saw widespread and effective service even so.

Many seaplanes were altogether more modern and impressive than the floatplanes, however. Britain and the USA

ARADO 196

The Arado 196 was designed to be used as a catapult-launched reconnaissance aircraft but also saw service operating from land bases. Over 500 were made and the first entered service a few weeks before the start of the war. Among other successes Arado 196s participated in the capture of the British submarine *Seal* in 1940.

SPEED: 311kph (193mph)
RANGE: 1,080km (670 miles)
ENGINE: 830hp Bramo 323
CREW: 2
ARMAMENT: 100kg (220lb) bombs, 2 x 20mm (0.79in) cannon + 2 machine-guns

converted the long-range civilian flying boats of Imperial Airways and Pan Am to military duties but in addition had purpose-built machines. Britain's was the Short Sunderland, with a patrol endurance of up to 16 hours and a weapons fit of depth charges, bombs or mines and many defensive machine-guns.

OTHER DESIGNS

The USA had two highly successful twin-engined designs: the Martin PBM Mariner and the Consolidated PBY Catalina (the Catalina in particular also served with numerous Allied countries). The Mariner was a slightly more capable design than the Catalina, with a longer range (5,600km/3,500 miles) and larger bomb capacity. However, the Catalina was produced in slightly larger numbers and was better known. The USA's four-engined type (the Consolidated PB2Y Coronado) was not as successful as its smaller stablemates.

By contrast Japan's main four-engined type was probably the best seaplane in use in WWII. The extended-range versions of the Kawanishi H8K "Mavis" could cover over 7,000km (4,350 miles) and were so heavily armed that they were very difficult to shoot down.

Even bigger was Germany's six-engined Blohm und Voss 222 Wiking, with an astonishing endurance of up to 28 hours; however, it was mainly used for transport duties.

LAND-BASED SUPPORT

Various nations employed smaller utility types for maritime patrol duties, including (in the early-war years) Britain's Avro

Above: A US PBY-5 Catalina on patrol off the Aleutian Islands in the North Pacific, early 1944.

Anson. It was very slow but highly reliable – engine failures far out to sea were never very popular with crews.

Effective longer-range patrol and anti-submarine operations came into their own with Anglo-American developments in air-to-surface radar. Variants of a number of large bombers were used for this role, most notably versions of the Consolidated B-24 Liberator. In addition there was the PB4Y-2 Privateer, a new type based on the Liberator and specifically designed for maritime strike and reconnaissance. It was larger than the B-24 and had a single not a twin tail. Several hundred saw service in 1944–5.

Small numbers of Germany's Focke-Wulf 200 Kondor, a converted airliner, played a significant role in the early stages of the Battle of the Atlantic. However, the Luftwaffe generally neglected maritime tasks – much to Germany's detriment.

SHORT SUNDERLAND

The Short Sunderland entered RAF service in 1938 and 749 were built before production ended in 1946. Several marks were produced with increased defensive armament and improved radar and other systems. As well as sinking U-boats, several Sunderlands fought off attacks by up to eight German aircraft.

SPEED: 336kph (210mph)
RANGE: 2,850km (1,770 miles)
ENGINES: 4 x 1,065hp Bristol Pegasus XVIII (in Mk 2)
CREW: 8–11
ARMAMENT: 900kg (2,000lb) bombs, 8 x 0.303in (7.7mm) machine-guns

The Central Pacific Campaign

From late 1943, the US Navy and Marines fought a new kind of war across the Central Pacific, characterized by the description "island-hopping", which relied as much on novel methods of logistical support as the efforts of the combat forces.

In 1943 the US Navy took 15 fleet or light aircraft carriers into service, but the Japanese commissioned only 1. The USA also out-produced Japan by 4 to 1 in submarines, roughly 10 to 1 in destroyers and by similar wide margins in every other class of munitions.

Nor was Japan husbanding its resources well. The already diminished cadre of trained carrier aircrews was further decimated in the early part of the year by being deployed on land in the Solomons and New Guinea. Training and improved

Right: The carrier USS *Lexington* refuelling at sea during the Tarawa operation, November 1943.

Below: US Marines lie dead amid the devastation and debris on the beach at Tarawa.

KEY FACTS

PLACE: Tarawa Atoll, Gilbert Islands

DATE: 20–3 November 1943

OUTCOME: Japanese garrison wiped out despite heavy American casualties.

---

equipment (notably better aircraft and radar) were also offsetting Japan's previous qualitative advantage in some areas, in air combat and naval night fighting, for example.

Even with potential combat superiority, the US forces still had the problem of distance to overcome: the US Pacific Fleet's base at Pearl Harbor was 3,200km (2,000 miles) from the West Coast; the Solomon Islands 14,500km (9,000 miles) from California; Australia still farther. To keep troops, ships and aircraft fed, fuelled and armed over such vast distances required a huge logistic effort.

The solution was a new type of naval organization, the fleet train, including tankers and other kinds of supply ship, so that combat vessels could be replenished at sea during assault operations far from any base. Then a range of new bases was improvised by rapidly building shore installations and installing floating docks in various previously tranquil Pacific lagoons – like Ulithi in the Caroline Islands (an important base from September 1944).

THE GILBERT ISLANDS
The first step was to make landings from 20 November 1943 on the Tarawa and Makin Atolls in the Gilbert Islands. The Makin landings, mainly on Butaritari Island, were fiercely contested, but the US Army troops wiped out all resistance by the 23rd.

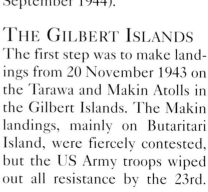

Above: US Marines ready to move inland from their landing beach, Eniwetok, February 1944.

However, Betio Island, the main target in the Tarawa Atoll, was a different matter.

Only 3km (2 miles) long and nowhere wider than 800m (880yd), none of Betio is more than 3m (9ft) above sea level. By November 1943 the 4,800-strong Japanese garrison had built a formidable network of bunkers and gun positions that largely survived the preliminary bombardments. The landing force, principally the 2nd US Marine Division, also lacked information on the depth of water in the lagoon and over the coral reef. For these reasons some 1,500 of the initial 5,000-man assault force became casualties on the first day, but the survivors held on. By the 23rd the only living Japanese were a handful of wounded, plus a few captured Korean labourers. US Marine casualties were over 1,000 dead and 2,000 wounded.

THE MARSHALL ISLANDS

The next stage was to seize control of the Marshall Islands. These were not as strongly fortified as was feared, although there were significant air forces deployed across a number of islands. These were gradually worn down from late 1943 by carrier attacks and land-based aircraft from the Gilberts.

On 31 January 1944 US troops began landings on Kwajalein, Roi-Namur and Majuro. The fiercest fighting was on Kwajalein, but by 4 February the 8,700-strong garrison had fought virtually to the last man, inflicting 370 dead on the attackers. Later in February Eniwetok was also captured. At the same time Japan's greatest

Above: Seabees setting up communications facilities on one of the Gilbert Islands.

overseas base, at Truk in the Caroline Islands, was heavily hit by the US carrier forces. The USA's relentless progress would continue without respite for the Japanese.

Landing Craft and Amphibious Vehicles

Amphibious vehicles and the many kinds of larger landing craft were vital to the Anglo-American war effort. Without them the whole counter-offensive in the Pacific and the defeat of Germany in western Europe would have been impossible.

Although amphibious operations had a long history, as late as a 1938 exercise (led by a certain Brigadier Montgomery) British troops were mainly using rowing boats to get ashore.

By then, however, the Japanese had a purpose-built 8,000-ton landing ship in service – the *Shinshu Maru* – which could deploy new Daihatsu landing craft from the stern. Japan subsequently introduced a small number of additional landing ships and landing craft, employing them successfully in its early offensive campaigns. In addition there were a few Toyota SUKI amphibious trucks in service, too.

US MARINE IDEAS

The other pioneer in landing operations was the US Marine Corps. In the inter-war period the Marines developed many of the ideas on amphibious warfare that would be put into action in WWII. In 1938 they began tests on so-called Higgins boats (in part copied from Daihatsu types seen in action in China). The wooden Higgins boats would be developed into more robust metal Landing Craft, Vehicle, personnel (LCVP), used by the

Below: US Army troops aboard an LCVP heading for Omaha Beach on D-Day. An LCVP typically carried 36 troops and 3 crew.

LCT MARK V

Several designs of Landing Craft Tank were built in Britain and the US from November 1940. The Mark V was a US design. This example is shown during operations on Rendova Island in the New Georgia group during 1943.

LENGTH: 35m (114ft 2in)
BEAM: 10m (32ft 8in)
DISPLACEMENT: 120 tonnes
LOAD: 127 tonnes cargo or
  5 M4 Sherman tanks
SPEED: 7 knots
ARMAMENT: 2 x 20mm (0.79in)
  AA guns
CREW: 13

thousand in many WWII campaigns. In 1941 the Marines also ordered the first of the important Landing Vehicle Tracked (LVT) series of amphibians. Later LVTs could carry 3 tonnes of cargo and be armed with weapons as large as a 75mm (2.95in) howitzer. Over 18,000 were built.

Even more prolifically produced was the DUKW amphibious truck. A final American amphibian was the smaller Studebaker M29 Weasel cargo carrier. All these amphibians were also used by the British.

BRITISH TYPES

By 1940 Britain had a small number of Landing Craft Assault (LCA) and Landing Craft Mechanized (LCM) in service – capable of carrying an infantry platoon or a tank, respectively – and soon added a number of other types. There was more than one variety of Landing Ship Infantry (LSI), converted either from ferries or small liners, and purpose-built Landing Ships Dock (LSD).

Rough American equivalents were Auxiliary Personnel Attack (APA) ships and Attack Cargo (AKA) ships. These various "ships" were not designed to reach all the way to landing beaches themselves but carried smaller "craft" into which they would load troops and supplies for the assault landings.

Also designed in Britain from 1940 – and subsequently made in large numbers in the USA – were various marks of Landing Craft Tank (LCT). The largest were around 56m (185ft) long and could carry nine Sherman tanks, landing them from a ramp at the bow.

These landing craft were among the smallest of a variety of types made for so-called shore-to-shore operations. They would be loaded in friendly territory and then sail under their own power to the landing area. Such vessels included Landing Ships Tank (LST) and Landing Craft Infantry, Large

Above: In the later stages of the Pacific war many Japanese landing craft, like this one on Saipan, were destroyed while trying to support island garrisons under attack.

(LCI/L); around a thousand of each were built. An LST could carry some 20 tanks and an LCI/L 180–210 infantrymen.

OTHER USES

Many landing vessels of all sizes were converted to specialized uses. Some became command or hospital vessels and others gave covering anti-aircraft fire. The most spectacular were the rocket-armed versions, which could fire a devastating barrage of up to 1,000 rockets into a landing area. By 1945 the variety and sophistication of landing craft and their tactics, and the weight of fire support available, meant that an assault landing could be made successfully against almost any opposition – hence the Japanese decision on Okinawa not to oppose the American landings at all.

DUKW

The DUKW amphibian was a 6 x 6 on land, based on a 2.5-tonne General Motors truck. It was used by the thousand in all theatres to unload transport ships and carry their cargoes inland to the troops. The examples shown are in British service preparing for D-Day.

LENGTH: 9.45m (31ft)
WIDTH: 2.49m (8ft 2in)
WEIGHT: 6.7 tonnes + 2.3 tonnes cargo
ENGINE: 91.5hp GMC
SPEED: 80kph (50mph) land; 10kph (6.2mph) water

The Marianas Campaign

The capture of the Marianas and Japan's crushing defeat in the naval Battle of the Philippine Sea ruptured the perimeter of its empire. The southern areas were being cut off as the war approached ever nearer to the Home Islands.

Although by mid-1944 the outer reaches of the Japanese Empire were increasingly helpless in the face of American advances, Japan's leaders still dreamed of setting all right by victory in a decisive naval battle. Their plan, code-named A-Go, was to mount a series of attacks by carrier and land-based aircraft to defeat the main American forces.

Unfortunately, not only did the American fleet have twice as many carrier aircraft (950:470)

Below: US troops in action on Saipan. The landings were made by the 2nd and 4th Marine Divisions and the 27th (Army) Infantry Division.

Above: Admiral Ozawa commanded Japanese naval forces in the Battle of the Philippine Sea and the carrier decoy force at Leyte Gulf.

but – in addition to the by now usual American advantages in code breaking and intelligence generally – the A-Go plan had been passed to them after being taken from a Japanese officer's crashed aircraft by guerrillas in the Philippines.

SAIPAN, TINIAN, GUAM

On 15 June two US Marine divisions landed on Saipan to set the land fighting in motion. Saipan had the strongest Japanese garrison in the area, some 27,000 men, but by 9 July the last effective resistance was destroyed. The landings on nearby Tinian began on 24 July; here the 6,200-strong garrison had been wiped out by 1 August. Guam was also captured from its 19,000-man garrison after fierce fighting, with the landings beginning on 21 July; organized Japanese resistance ceased by 10 August.

The Japanese garrisons all fought virtually to the last man. The American forces, mainly US Marines, lost some 5,000 dead. By then, however, the A-Go battle had long since been fought.

BATTLE OF THE PHILIPPINE SEA

When the preliminary American air attacks on the islands started, Admiral Toyoda Soemu, Commander of the Japanese Combined Fleet, gave the order for A-Go to begin. Almost immediately things went wrong

Above: The carrier *Zuikaku* and several destroyers trying to dodge US air attacks, 20 July 1944. *Zuikaku* was hit but not sunk.

for the Japanese. As the fleet approached it was spotted by US submarines. At the same time American air attacks destroyed most of the Japanese aircraft force on the Mariana Islands themselves, as well as those on Iwo Jima and other islands that were meant to be sent to the Marianas. Crucially, the local commander on the Marianas hid the bad news from the fleet's tactical commander, Admiral Ozawa Jisaburo, who believed that the American forces were being worn down as he approached.

The one remaining Japanese advantage was in the range of both their combat and reconnaissance aircraft. It enabled them to find the American carriers and send off their strikes first, in the morning of 19 June. In all Ozawa sent off 4 strikes involving some 370 aircraft in the course of the day; about two-thirds were shot down while other Japanese

planes were destroyed over Guam. Only one American ship was hit by a bomb and only 29 American aircraft lost.

Japanese aircraft and their hastily trained crews could no longer compete with the American pilots and anti-aircraft gunners, who called the day's events the "Great Marianas Turkey Shoot". Even worse for the Japanese, two of their largest carriers – the *Taiho* and *Shokaku* – were torpedoed and sunk by American submarines.

Despite all this, Ozawa tried to continue the battle, believing the optimistic reports that were still reaching him. On 20 June American reconnaissance did not fix the Japanese position until well into the afternoon when some 130 bombers and 85 fighters were sent to attack, even though the American commanders knew that their pilots would have to return to the carriers after dark. Another Japanese carrier was sunk and 3 more damaged for the loss of 20 American aircraft. Over 70 craft were lost while returning to their carriers, but most of the crews were picked up.

By contrast hardly any crew members of the more than 400 Japanese aircraft lost in the battle were saved. Thus, the biggest ever carrier battle also marked the effective demise of the Japanese carrier force. They still had ships but hardly any trained crews to fly from them.

ADMIRAL CHESTER NIMITZ

Nimitz (1885–1966) commanded the US Pacific Fleet from a few days after Pearl Harbor until the end of the war. His leadership, more than that of any other commander, was responsible for the series of crushing defeats inflicted on Japan and the transformation of the US Navy from the low ebb of December 1941 to absolute dominance in 1944–5. His leadership was probably best seen in the difficult times of 1942, when his able use of intelligence led directly to the first great success at Midway.

Above: Fleet Admiral Chester Nimitz pictured in 1945.

Naval Fighters

Like their land-based counterparts, fighters designed for operations from aircraft carriers had to be fast, manoeuvrable and well armed, but long range and robust construction for operations in difficult maritime conditions were equally valuable.

Since only Britain, the USA and Japan operated aircraft carriers, these were the only nations that had aircraft in this class, though on occasion the aircraft concerned also operated from land bases.

BRITISH DESIGNS

Britain's early-war naval fighters were particularly poor. In 1939 Britain's Fleet Air Arm still employed the biplane Sea

Below: Grumman Hellcats leading an air group preparing to take off from the carrier *Lexington* in 1944.

Gladiator type and also had the turret-armed Blackburn Roc, which (though a monoplane) was even slower than the Gladiator. These were replaced from 1940 by the Fairey Fulmar, which fought reasonably effectively against Italian aircraft in the Mediterranean but did not compare well with other nations' designs. The later Fairey Firefly, also a relatively large two-seat aircraft, had performance of a more modern standard, with a top speed of 508kph (316mph), and could also carry a useful bomb load.

The best home-made pure fighters employed afloat by the Royal Navy were conversions of the Spitfire, known in naval service as the Seafire. Various marks of Seafire were used and, like the parent design, were fast, manoeuvrable and well armed – but lacking in range. In the later-war years most naval fighters in British service were the US types described below.

THE JAPANESE NAVY

The best-known Japanese naval fighters came from the Mitsubishi company. The A5M

Above: A formation of US Navy Grumman F4F Wildcats in flight.

Above: Like other two-seat fighters, the Fairey Fulmar did not fare well in combat with modern single-seat designs, despite its heavy armament.

"Claude" served in China in the late 1930s and to some extent in the early part of the Pacific War. With a top speed of some 435kph (270mph), it was surprisingly fast for a fixed-undercarriage design and highly manoeuvrable.

The next Mitsubishi fighter – the A6M Type 0 "Zeke" (or "Zero") – was truly remarkable. In service from 1940, it had unequalled combat manoeuvrability at that time, was well armed with 2 x 20mm (0.79in) cannon and 2 machine-guns and had an astonishing 950km (600 mile) radius of action. It out-classed all Allied opponents until at least late 1942. The more powerful late-war A6M5 variant was the most-produced Zero. However, by then the Allies had well-trained pilots in abundance and aircraft with the heavy armament and perform-ance to exploit the Zero's weak-nesses of light construction and lack of armour for the pilot.

Other notable Japanese Navy fighters were various models of the Mitsubishi J2M Raiden "Jack" and the Kawanishi N1K Shiden "George". These mostly served in land-based roles, latterly in defence of the Home Islands against American B-29 bombing raids.

AMERICAN TYPES

The US Navy also began the war with a soon-to-be-phased-out pre-war design, the Brewster F2A Buffalo, the US Navy's first monoplane fighter. This served at Midway and with the British RAF in Malaya in 1941–2 but was clearly no match for the Zero.

Already replacing it by then was the Grumman F4F Wildcat (also known as the Martlet in British service). Updated variants of the Wildcat re-mained in service until 1945. By then it had been supple-mented by a larger and more powerful Grumman design – the F6F Hellcat – which some commentators describe as the best carrier fighter of the war. It was highly manoeuvrable and extremely robust, an advan-tage not just in combat but also in the common occurrence of heavy carrier landings. With a top speed of 620kph (385mph), it had more than adequate performance.

Challenging the Hellcat was the US Navy's other main late-war fighter – the Chance Vought

MITSUBISHI A6M2 ZERO

In 1941–2 the Zero fighter seemed invincible, with better performance than any Allied aircraft. However, it was very lightly built with at first no cockpit armour or self-sealing fuel tanks. These were introduced in later models in use until 1945.

CREW: 1
SPEED: 533kph (331mph)
RANGE: 3,100km (1,930 miles) with drop tank
ENGINE: 940hp Nakajima NK1C Sakae radial
ARMAMENT: 2 x 20mm (0.79in) cannon + 2 x 7.7mm (0.303in) machine-guns

F4U Corsair (in service from October 1942). Significantly faster than the Hellcat – over 700kph (435mph) in late-war variants – the F4U was also used effectively as a fighter-bomber.

Submarines and Bombers

For months before its surrender Japan's empire had been made useless to the mother country by American submarine attacks, while virtually every major city in the Home Islands was flattened in the American bombing campaign of 1944–5.

Japan's plans to expand its empire and gain access to the natural resources that the Home Islands lacked had been maturing for years before the war. But remarkably little thought had been given to protecting Japan's links with the empire or to moving the empire's products safely. If not perhaps a cause of Japan's defeat, this glaring oversight was certainly a large part of the reason why the USA won the war so quickly.

After Pearl Harbor the only part of the US Pacific Fleet that could immediately attack was the submarine force but, like the Germans in 1939–40, their torpedoes were very poor, often failing to explode or going off

Below: Emperor Hirohito (centre) inspects the scenes of devastation in Tokyo, March 1945. About a quarter of the city was destroyed.

Above: General Curtis LeMay commanded the US bomber forces in the Marianas from January 1945.

course. The problems were not fully solved until mid-1943. From then until the end of the war American submarines took an ever greater toll of Japan's shipping. Although Japan depended on imports to keep its economy operating and needed to send troops, supplies and

arms to all its outposts, the Japanese Navy had neglected anti-submarine operations. For example, few ships had underwater sensors even as late as 1942, nor were there plans for a convoy system.

The US code-breaking and intelligence service often guided submarines to the right places; once there, they benefitted from radar to find their targets and warn of air attacks.

CRIPPLING LOSSES

By the end of the war, almost 5 million tonnes of shipping, along with numerous warships, had been sunk directly by submarines and over 2 million tons more by mines, both air- and

Below: As well as attacking enemy forces US submarines often rescued downed airmen, like these aboard the USS *Tang* in 1944.

Below: A sinking Japanese merchant ship seen through the periscope of the USS *Wahoo*, 1943.

Right: Admiral Charles Lockwood (left), the very able commander of US Pacific Fleet submarines, aboard the USS *Balao* in 1945.

submarine-laid. By the summer of 1945 American submarines were even operating in the land-locked Sea of Japan, cutting the links between Japan and its large armies on the Asian mainland. Japan was in effect now totally blockaded.

RUIN FROM THE AIR

Unlike the submarine attacks, the bombing of Japan could only start when the long-range Boeing B-29 Superfortress aircraft became available. The campaign began on 15 June 1944 when 50 B-29s based in India flew via airfields in China to attack targets on Kyushu, the southernmost of the main Japanese islands. A further 50 or so attacks were flown using the Chinese bases before Japanese Army advances over-ran them in early 1945, but only a handful struck targets in Japan itself.

By November 1944 new bases in the Marianas had been opened from which virtually the whole of Japan could be hit. At first, US tactics were to make high-altitude daylight raids, attempting precision bombing of aircraft factories and similar targets. These met determined Japanese air defences and achieved disappointing results.

Early in 1945 experiments with a new technique were begun – low-level incendiary bombing of cities by night. This had several potential advantages: if attacking civilians directly was acceptable, then Japan's cities, with buildings predominately made of wood and paper, were vulnerable; Japan had little night-fighting capability so losses would be lower; and bombers could carry heavier loads and yet suffer fewer mechanical problems because of the lower altitudes.

After a number of trial operations, the new tactic was put into action in earnest on the night of 9–10 March 1945, when some 280 B-29s attacked Tokyo. The incendiaries raised a huge firestorm that left perhaps 120,000 Japanese civilians dead, making it the most devastating air raid of the war (not excluding the atomic attacks yet to come). In succeeding weeks city after city suffered a similar fate, with the attacks being supported from late April by escort fighters flying from airfields on newly captured Iwo Jima.

By late July the bombers were running out of large towns to target, Japan's economy was in ruins, at least 800,000 were dead and up to 10 million homeless. Perhaps most important of all, the Emperor and some of Japan's civilian leaders were becoming convinced that the war would have to be ended.

Submarine Classes

WWII submarines had limited capabilities, but no one doubted that they were potentially war-winning weapons. Germany's 1,000 U-boats failed in the Battle of the Atlantic, but the US Navy's submarines fatally weakened Japan.

Germany's U-boat force depended for most of the war on two main designs: the smaller (750 tons surfaced) Type VII and larger (1,000 tons) Type IX. These designs, clearly derived from WWI U-boats, were both well-engineered and robustly built for deep diving.

U-BOAT DEVELOPMENTS

Ten Type XIV supply U-boats were also built. These played an important role in extending the operational range, especially of the Type VIIs, but all the Type XIVs were hunted down as a priority by the Allied forces using code-breaking information. By the mid-war years the U-boats were outmatched by

Below: The Italian Tritone-class boat *Marea* off Bermuda in 1944. *Marea* was then being used by the Allies for anti-submarine training.

Allied anti-submarine forces, so work on new technologies to overcome this was stepped up. First introduced was a breathing tube, or *Schnorchel* (a pre-war Dutch invention), designed to enable the submarine to run its main engines while submerged and difficult to detect. This worked up to a point but had various disadvantages when in use. More promising was work to streamline submarine hulls

Above: The Gato-class USS *Barb*, seen in San Francisco Bay in May 1945, returning to action after a refit.

and step up battery capacity. A few Type XXI and XXIII U-boats using this technology came into service shortly before the end of the war – their high underwater speed made them very difficult to counter. More might have been built if Germany had not wasted much

effort on the abortive development of the Walther system, which used hydrogen peroxide to provide oxygen so that the main engines could run when the boat was submerged.

EUROPEAN CLASSES

Britain had three main classes of submarine during the war: U, S and T (in ascending order of size). The 550-ton U-class boats were designed for training duties but, in the event, were used effectively in action in the confined Mediterranean waters. All the British boats had the merit of being fast-diving and carrying a heavy armament of bow torpedo tubes – eight in the T class compared to six or even four in other nations' boats.

Submarines built for Pacific service (including the British T class) tended to be larger than those designed for European waters. In 1939 the largest submarine in service was France's *Surcouf* – 3,250 tons, armed with a twin 203mm (7.99in) turret and carrying a floatplane.

JAPANESE SUBMARINES

Japan's wartime I-400 class (three built) were even bigger, at 5,200 tons, and could carry three aircraft, intended to attack the locks on the Panama Canal. Other nations experimented with monster submarines before WWII, but these were the only examples to see any service.

Japan's standard submarines were unremarkable: relatively slow-diving and unable to dive very deep. Their advantage was using a 533mm (21in) version of the famous "Long Lance" torpedo, by far the best submarine torpedo of the war. Japanese tactics also emphasized attacks on

HMS *TUDOR*

The British T-class submarine *Tudor* was commissioned in 1944 and served against Japan to the end of the war, sinking ten ships. In all, 53 boats of the T class were completed, several of them serving with the Dutch Navy.

DISPLACEMENT: 1,290 tons surfaced; 1,560 tons submerged
LENGTH: 84.3m (276.5ft)
SURFACE SPEED: 15.5 knots
SUBMERGED SPEED: 9 knots
ARMAMENT: 11 x 21in (533mm) torpedo tubes + 1 x 4in (102mm) gun

enemy warships and disregarded attacks on supply ships. Though Japanese submarines sank the carriers *Wasp* and *Yorktown* among other successes, their contribution was limited. The largest-sized class was the 2,200-ton I-15 type.

US NAVY TYPES

American fleet submarines were all of high quality. The similar Gato, Balao and Tench classes saw much service. They were all roughly 1,500 tons and well designed, both in terms of radar and sonar equipment, as well as incidentals like air conditioning that helped make long Pacific patrols more comfortable for the crews. Unfortunately, for more than a year after Pearl Harbor, their torpedoes were very poor. When this fault was rectified US submarines practically wiped out the Japanese merchant fleet and sank many of the naval vessels sent to hunt them down.

Below: A U-boat after its surrender in 1945. Note the heavy anti-aircraft armament carried by most U-boats by this time.

Burma, 1943–4

The retreat in Burma in 1942 was the longest in British history, but by mid-1944 the British–Indian Fourteenth Army had inflicted what was then Japan's worst-ever defeat on land in the Battles of Imphal and Kohima.

Japan's aim in attacking Burma in 1942 was to cut the vital Burma Road supply route to China and, in due course, to advance onward to India. British intentions were to defend India and in time recover Burma and later Malaya, while US leaders were less interested in victory in Burma for its own sake than as a means of reopening a supply route to China and the development there of powerful attacks on Japan. In the end, though Japan's advance on India was thrown back in 1944, Burma's position far down the Allied priority list helped ensure neither British nor American plans would come fully to fruition.

Burma's monsoon season made large-scale military operations impossible each year from

Above: Orde Wingate led the Chindits in 1943, but died in 1944 during a second Chindit operation.

late May to early November. By later in 1942 Allied forces had been rebuilt sufficiently after the debacle earlier in the year to attempt a modest offensive in the coastal Arakan region. After several months of fighting, the Japanese recaptured most of the ground briefly gained – an outcome which confirmed that they still held the upper hand tactically and psychologically in jungle battle.

THE CHINDITS

The only other offensive move possible in that period was an entirely novel "long-range penetration" operation by a force called the Chindits, led by the charismatic (and barely sane) Brigadier Orde Wingate. Supplied from the air, they operated well behind Japanese lines, cutting communications and attacking bases for some weeks before being ordered to return to India. Wingate was lionized by Churchill; he and his men were celebrated in the press as having shown that the Japanese were not invincible in the jungle. Reality was different: a third of the force was lost, hardly any of the survivors were fit for future operations and their attacks achieved nothing.

The real way forward was shown by another series of Arakan battles in late 1943– early 1944. The British advance there was intended as part of a general Allied offensive, while the Japanese planned their own attack as a preliminary to a drive into India in the Imphal area, far inland to the north-east. In the event part of the British force was surrounded for a time, a circumstance that had often led to

Below: A British 25pdr (88mm) gun position in the Burmese jungle.

Right: Japanese and Indian National Army men during the Imphal campaign.

defeat in the past. Now better-trained, the troops stood their ground (with the benefit of air supply) and simply out-fought the Japanese. Much of this success was down to General Bill Slim, in command of the British forces in Burma, Fourteenth Army, since October 1943.

IMPHAL AND KOHIMA

A similar pattern emerged on the main inland battle front, after much hard fighting. Under the codename U-Go, the Japanese planned to advance into India with attacks led by General Mutaguchi Renya's Fifteenth Army. Slim and other British commanders expected a Japanese attack but were caught by surprise by the speed of their advance. As a result a small force was surrounded at Kohima and a larger one at Imphal on 4–5 April. Once again the plan was to hold firm with air support until relief troops fought their way through from the north.

Right: Japanese and Indian National Army men during the Imphal campaign.

The siege of Kohima was broken by late April after a vicious close-quarter battle in and around the town. It took until well into June for the advance to link up with the Imphal garrison and another month before the Japanese finally retreated. While Slim's men had been receiving thousands of tons of supplies and substantial reinforcements. by

Below: British troops serving with the 19th Indian Division in action with 3in (76.2mm) mortars near Imphal.

air throughout the battle, the Japanese had been operating on a logistical shoestring; many were now starving. They lost about 60,000 casualties, more than half of them dead, compared to some 17,000 British and Indian killed and wounded.

THE INDIAN NATIONAL ARMY

This force, recruited mainly from Indian Army prisoners of war captured in Malaya in 1942, was formed to support the cause of Indian independence by fighting with the Japanese against the British. It had various changes of leaders and organization but from mid-1943 was led by Subhas Chandra Bose and had a maximum strength of about 20,000. It was never fully trusted by the Japanese and was mainly used in small detachments. INA troops fought (but usually not very effectively) in various battles in Burma from early 1944 to the end of the war.

The Battle of Leyte Gulf

The three separate engagements comprising the Battle of Leyte Gulf together make up the largest naval battle in history. Despite their huge superiority, the US forces only just escaped terrible losses to the Leyte invasion fleet.

As the US advance gathered pace in mid-1944, commanders argued about how to attack Japan. General MacArthur felt the Philippines should be recaptured, but US Navy leaders favoured Formosa (now Taiwan) as the next target. In the end, landings on Leyte were planned for 20 October, with the main Philippine island of Luzon to be attacked at the end of the year.

THE JAPANESE PLAN

Japan still had many big-gun ships, including the *Yamato* and *Musashi*, the largest battleships ever built. Japan's admirals also

KEY FACTS

PLACE: Three separate battles among and near the Philippine Islands

DATE: 24–5 October 1944

OUTCOME: Americans narrowly avoided a disaster to win a major victory.

THE BATTLE OF LEYTE GULF
The Japanese approach and the three Leyte Gulf battles, within the Philippine Islands complex.

wanted an all-out battle in which these ships might turn the tide of the war. The Japanese naval air arm was no longer powerful: it had aircraft carriers but few trained crews or planes. Admiral Toyoda's Sho-Go plan (Operation Victory) called for carriers to divert the main American strength to the north, with most of the heavy ships attacking the vulnerable US landing fleet from the west through various channels between the Philippine Islands.

The US landings on Leyte began on 20 October. The main Japanese force, Force A, sailed from Borneo on the 22nd. US submarines sank two cruisers and reported Force A's move the next day. On 24 October the American carrier *Princeton* was sunk by land-based aircraft.

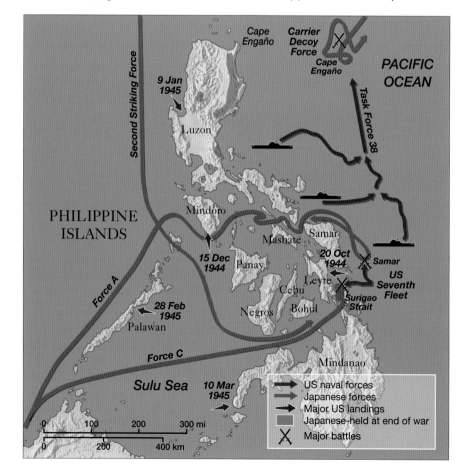

Above: Admiral Thomas Kinkaid (left) and General Walter Krueger, commander of Sixth Army on Leyte.

Above: The carrier USS *Princeton* on fire and sinking, 24 October.

Force A was attacked by US carrier aircraft in the Sibuyan Sea; the *Musashi* was sunk by bombs and torpedoes. Admiral Kurita Takeo ordered a brief turn away because of these attacks.

HALSEY'S MISTAKE

The two main US naval forces were Admiral William Halsey's Third Fleet, including the main carrier force, Task Force 38; and Admiral Thomas Kinkaid's Seventh Fleet of transport and support vessels. Halsey now assumed (incorrectly) that Force A had retreated for good and (correctly) that Kinkaid's shore-bombardment ships could cope with the Japanese southern force. He therefore felt free to take his main force north against the Japanese carriers.

Thus, in the early hours of the 25th, the old battleships of the Seventh Fleet triumphed in a night gun and torpedo battle in the Surigao Strait; two Japanese battleships were among the vessels sunk. Now, after earlier abortive strikes, the Japanese carriers in the north had barely a couple of dozen aircraft left between them to face the ten full-strength carriers of TF 38. Three Japanese carriers and

Below: The destroyer *Akizuki*, sunk during the Battle of Cape Engaño.

several other ships were sunk during the 25th in the Battle of Cape Engaño.

More would have been sent down but part of the US attack had to be called off by emergency messages from the fleet off Leyte, suddenly facing Force A. This should easily have had the gun-power to brush aside the weak escort carriers and destroyers that stood between it and the virtually defenceless invasion transports.

Instead resolute American defence and a developing fuel shortage persuaded Kurita to turn away feebly after an inconclusive combat, known as the Battle of Samar. The various Japanese forces suffered further losses during their retreats. When all were counted, what might have been an American disaster thus turned into another catastrophe for Japan's fleet.

KAMIKAZE ATTACKS

By 1944 numerous Japanese garrisons had made desperate suicidal attacks as a last resort but suicide kamikaze attacks, from October 1944, became a deliberate first-choice tactic. The first organized kamikaze attack was on 21 October off Leyte; the cruiser HMAS *Australia* was badly damaged. The first kamikazes used standard fighters, but aircraft fitted with heavy bombs were later employed along with various purpose-built craft, including piloted flying bombs, explosive motor boats, manned torpedoes and midget submarines. Around 500 kamikaze air attacks were made during the Philippine campaign and perhaps 2,000 during the fighting on Okinawa in 1945.

Above: Kamikaze pilots are given their ceremonial headbands.

Naval Bombers and Torpedo Aircraft

An effective air strike by a naval force usually employed a mixture of dive-bombers and torpedo-carrying aircraft. Aircraft in both these roles could be desperately vulnerable to fighters and anti-aircraft fire; casualties were often heavy.

As in other categories of maritime aircraft, the only countries with designs of these types were Britain, Japan and the USA. In 1939 the British aircraft in service were already outdated and later designs were little better. However, for the Japanese, aircraft types that were effective at first – especially when backed by a superior naval fighter – were not adequately replaced when faced with ever stronger and more sophisticated American forces.

THE ROYAL NAVY

Britain's main torpedo-bomber in 1939 was the ancient-looking Fairey Swordfish biplane, slow and with a range of only 880km (550 miles) on a full load. Remarkably, fitted with radar

Below: Nakajima B5N "Kate" torpedo-bombers in flight over a Japanese fleet base before the war.

and carrying depth charges and rockets, Swordfish were still in use in the anti-submarine role in 1945. By this time the Swordfish's replacement, the biplane Fairey Albacore, had already been retired. The early-war dive-bomber, the monoplane Blackburn Skua, was also poor: slow and with a bomb load of only 226kg (500lb). From 1943 a further Fairey design – the Barracuda (this time finally a monoplane) – entered service. It was more often used as a bomber, though it had been designed as a torpedo aircraft.

AMERICAN DESIGNS

The US Navy's early-war aircraft were both from the Douglas company. The SBD Dauntless was the dive-bomber. Its crews claimed that its designation stood for "Slow But Deadly". In battles like Midway it was indeed very successful.

CURTISS SB2C HELLDIVER

First flown in 1940, the Curtiss Helldiver made its operational debut in an attack on Rabaul in November 1943. In all 7,140 examples were built. It served with several Allied countries through 1945.

SPEED: 473kph (294mph)
RANGE: 1,900km (1,200 miles)
CREW: 2
ENGINE: 1,900hp Wright R-2600 Cyclone radial
ARMAMENT: 900kg (2,000lb) bombs, half in internal bomb bay; 2 x 0.79in (20mm) cannon + 2 x 0.3in (7.62mm) machine-guns

It was used by some units until 1945. Its early-war companion was the TBD Devastator, which was not so well regarded: slow and under-powered and obviously very vulnerable.

The Dauntless was replaced in front-line carrier service from late 1943 by the Curtiss SB2C Helldiver. This had good performance figures, but also

had reliability problems; it was not well liked by its crews. The replacement for the Devastator was more successful.

Designed and originally produced as the TBF by Grumman – and later made by Eastern Aircraft as the TBM – the Avenger was a robust and capable aircraft. Conceived only in 1940, and rushed into action in mid-1942, it served successfully for the remainder of the war. It more often carried bombs or even air-to-ground rockets than the torpedo for which it had been designed.

JAPAN'S REPLY

The attack on Pearl Harbor by Japan was carried out by aircraft that became known to the Allies as the "Val" and "Kate". The Aichi D3A "Val" dive-bomber could carry only the relatively modest bomb load of 360kg (800lb) but was surprisingly fast, at 397kph (247mph), for an aircraft with a fixed undercarriage and was credited with particular accuracy when dive-bombing.

It was replaced by the Yokosuka D4Y "Judy", which was produced in a number of

sub-types, but none of them was satisfactory. A further Aichi design, the B7A "Grace", was potentially better but it only entered service in 1944.

With the "Val" in 1941 was the Nakajima B5N "Kate". Again it performed well in the early months of war but it, too, dated back to the later 1930s and was due for replacement. The new type, the Nakajima

Above: A Douglas Dauntless over New Guinea in 1944. Almost 6,000 of the type were built.

B6N "Jill", arrived only in June 1944. It would have been an effective aircraft, with good range and ordnance-carrying capabilities, but unfortunately the Japanese Navy no longer had any trained pilots and aircraft carriers for it to fly from.

FAIREY SWORDFISH MARK 1

Although it had first flown in 1934, the "Stringbag" was still in widespread use in 1945. It played a vital role in many successful battles, including Taranto and the sinking of the *Bismarck*. In all 2,392 were built.

SPEED: 222kph (138mph)
RANGE: 880km (550 miles)
CREW: 3
ENGINE: 690hp Bristol Pegasus III radial
ARMAMENT: 1 x 18in (457mm) torpedo or 680kg (1,500lb) bombs; 2 x 0.303in (7.7mm) machine-guns

Conquest of the Philippines

When General MacArthur had left the Philippines in 1942, he had pledged: "I shall return." He landed on Leyte to fulfil that pledge in October 1944, but Japanese forces were still holding out on Luzon when the war ended almost a year later.

The planned Allied landings on Leyte in the south-west of the Philippine archipelago were hurriedly brought forward when aircraft carrier attacks suggested that the island might be relatively lightly defended. This proved to be a false hope, but in the end the established pattern of Pacific engagements was repeated, with the American forces gradually wearing down a tough and determined defending force.

THE BATTLE FOR LEYTE

General Walter Krueger's Sixth Army landed some 130,000 troops on the east coast of Leyte

Below: The material strength Japan could never approach. Massed US LSTs land supplies on Leyte.

Above: General MacArthur (in sunglasses) wades ashore on Leyte on his return to the Philippines.

on 20 October 1944, the first day of the operation. General MacArthur was also present, dramatically wading ashore from a landing craft (and then typically repeating the process

for the benefit of the assembled press cameramen). The initial Japanese force on Leyte was some 20,000 strong but General Yamashita, recently appointed as Japan's top commander in the Philippines, sent some 50,000 reinforcements to the island, though these took significant casualties and lost much equipment in transit.

Although the Japanese Navy was in the end decisively defeated in its attempts to intervene in the Battle of Leyte Gulf, the fighting on land was bitterly contested. Ultimately the Japanese had no answer to the American land and air firepower and organized resistance came to an end by late December. Casualties were as disproportionate as ever – over 60,000 Japanese dead compared to 3,600 Americans.

By then the US Eighth Army, commanded by General Robert Eichelberger, was also involved in the fight, having begun landings on the lightly defended Mindoro Island on the 15th of that month. As usual, they quickly built airfields to support future operations.

LIBERATING LUZON

General Yamashita had about a quarter of a million men left on Luzon by the end of the year, but inevitably they were greatly dispersed across the island's substantial area and were not well armed or equipped. Therefore, Yamashita decided not to

Above: A US M10 Tank Destroyer in action in the infantry-support role on Leyte in late 1944.

Above: An American casualty is taken for treatment amid the wreckage of Manila, February 1945.

contest in any great strength whatever landing beaches the Americans selected, but rather to retire gradually into the mountainous interior and hold out there for as long as possible.

On 9 January 1945 General MacArthur sent Sixth Army ashore in Lingayen Gulf. Kamikaze and conventional air attacks caused some damage to the landing fleet en route, but soon the few surviving Japanese aircraft were flown out to Formosa (Taiwan) and elsewhere.

TOWARD MANILA

As the Japanese invaders had in 1942, the US forces soon pushed south toward Manila, the capital. Here, in an exception to Yamashita's overall plan, the Japanese forces chose to stand and fight. The battle for

Right: A battery of 105mm (4.13in) howitzers shelling the Intramuros area of Manila in February 1945.

Manila lasted for a month, from early February to 3 March. Although MacArthur forbade air attacks, what the artillery bombardments and ground combat did not destroy, the Japanese blew up or set on fire. The city was flattened and about 1,000 Americans and 16,000 Japanese died in the battle along with probably 100,000 local civilians.

Operations on Luzon and elsewhere in the Philippines continued for the rest of the war, with organized Japanese forces gradually being confined in ever smaller and more remote areas. Filipino guerrillas, who had been active to some extent during the Japanese occupation, played an increasingly important role alongside the Americans in these operations. However, Yamashita still had 50,000 men under his command when he surrendered in August.

China's War

China lost more casualties in the war than any other nation apart from the Soviet Union, and the war between China and Japan was the longest of the series of conflicts that made up WWII.

The war in China was not a simple contest between two sides, as it was in most other areas of the conflict. By the late 1930s even as the Chinese Nationalist government of Jiang Jieshi (or Chiang Kai-shek in the then usual English spelling) was being pushed out of large areas of northern and coastal China, it was also coming into conflict with the growing power of the Chinese Communists, who were led by Mao Zedong (Mao Tse-tung).

INTERNAL STRUGGLES

In addition to these forces, and the million and a half Japanese troops in China and Manchuria by late 1941, were the significant armies of the puppet regimes set up by the Japanese and various local groupings of shifting loyalty. After Japan's

Above: Chinese Nationalist troops like these fought in Burma in 1944–5 to open the Ledo Road supply route.

attack in 1937, there was an official truce between the Nationalists and Communists, but in fact there was continued fighting between their forces at a local level. In the period up to December 1941 China's Communists were also successful in greatly expanding their organization in supposedly Japanese-controlled areas.

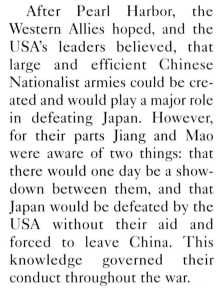

After Pearl Harbor, the Western Allies hoped, and the USA's leaders believed, that large and efficient Chinese Nationalist armies could be created and would play a major role in defeating Japan. However, for their parts Jiang and Mao were aware of two things: that there would one day be a showdown between them, and that Japan would be defeated by the USA without their aid and forced to leave China. This knowledge governed their conduct throughout the war.

The Communists would be more active in opposing the Japanese: in part for the credit they knew would increase their popularity among the Chinese people; and in part by default because the areas where they were building up their strength were also the areas where the Japanese were most active.

Jiang's position at the head of the Nationalist movement depended on his personal control over the army. Removing corruption and professionalizing it (as his American advisers, led by General Joseph Stilwell, wanted) would have jeopardized this control, so it was not done. Instead of being employed against the Japanese, American supplies were largely used, or held for later use, against the Communists.

By 1941 Japan's advances in China had mostly come to a halt. From then until 1944 the Japanese made few significant

Below: Mao Zedong inspecting men of the Communist Eighth Route Army in Shensi province in 1944.

forward moves. Although Japan had forces deployed across large swathes of China, they did not actually hold much more than the major cities and the links between them. They fought repeated "anti-bandit" campaigns in the countryside, plundering and destroying as they went. During the war years the Japanese-controlled areas came under increasing Communist influence – most of the "bandits" were in fact Communist guerrillas.

JAPANESE OFFENSIVES

There was one significant Japanese offensive in 1942 after some of the Doolittle raiders had flown on to China following their attack on Tokyo. Various Chinese and American air bases were captured and in all perhaps a quarter of a million

Below: Chongqing, the Chinese Nationalists' wartime capital, after a Japanese air attack in May 1939.

Chinese were killed, a significant proportion of them by biological weapons.

By 1944 the USA had built up its Fourteenth Air Force in China. Its attacks within China (and the first raids on Japan itself that followed) provoked a major Japanese attack, notably in south China – the Ichi-Go Offensive from April that year. The Nationalist forces facing

Above: A US C-46 flying supplies on the difficult "Over the Hump" route from India to China.

this advance collapsed and the American air bases were easily overrun. The Chinese did regain some ground in the far south in 1945, but by then no one believed that China had much part to play in finally defeating Japan.

Victory in Burma, 1944–5

Fourteenth Army's series of brilliant victories continued from late 1944 to the end of the war. The successes at Meiktila and Mandalay ensured that Burma would be almost completely liberated by May 1945.

Although any Japanese thoughts of invading India had been crushed at Imphal in the first half of 1944, the Allied aims of opening a road to China and recovering Burma were still to be fulfilled. British, Indian and other British Empire troops of Fourteenth Army would make the main advance into central Burma, as well as a secondary move along the coast. At the same time a combination of Chinese, American and British troops fought farther north and inland to create the land supply route to China.

THE LEDO ROAD

US leaders, including General Joseph Stilwell (the USA's top soldier in China until October 1944) believed that the Chinese Nationalists could play a major role in defeating Japan, if only they could be adequately equipped by the Allies. Although great efforts had been made to fly supplies "Over the Hump" of the mountain ranges between India and China, truly significant quantities could only be delivered by land.

The route chosen started at Ledo, the north-eastern end of India's rail system, and then ran southward into Japanese-held territory to link with the old Burma Road, which had supplied China before Pearl Harbor. Construction began in late 1942, accompanied by advances to clear the Japanese out of the way. The road was finally opened in January 1945.

Below: British infantrymen watch Indian Army Sherman tanks approaching Meiktila, March 1945.

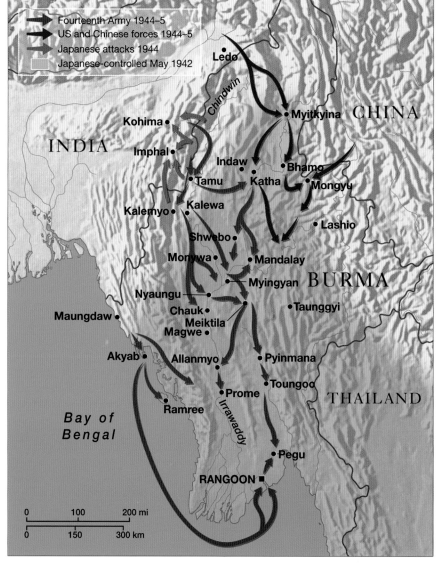

BURMA CAMPAIGN, 1943–5
Japan's defeats at Imphal and Kohima and the Allied offensive to liberate Burma.

Above: British troops crossing the River Chindwin at the start of their advance in late 1944.

The main battles in this sector were fought from late 1943 to August 1944, with the fighting being especially fierce around the town of Myitkyina. The Allied forces involved included the so-called Chinese Army in India, an American formation known as Merrill's Marauders and a second Chindit expedition that was much larger than the first.

SLIM'S MASTERPIECE

British leaders had a far more jaundiced (and as it turned out more realistic) view of what the Chinese Nationalists might contribute to the Allied war effort, preferring instead to advance in central Burma and plan seaborne operations farther south later.

Although the Japanese had substantially rebuilt their forces in Burma after the Imphal disaster, they were now clearly outmatched. The men of Fourteenth Army described themselves as the "forgotten army" because of their position

far from home and well down the Allied priority list. However, they were in no doubt that the quality of their training, tactics and weaponry made them more than the equal of the Japanese in combat. They also had lavish air support and, above all, in General Slim, a resourceful and totally trusted commander.

When the advance began in late 1944, a deception scheme helped convince the Japanese that Mandalay (the traditional capital of Burma) was the Allied target. Instead General Slim's main plan was to capture the vital communications centre of Meiktila to the south, which was taken in early March. The force at Meiktila was isolated for several weeks by Japanese counter-attacks, but by late March these had been defeated – and the relentless pressure from other parts of Fourteenth Army had also taken Mandalay.

The Japanese were now in all-out retreat, though still capable of fighting tough rearguard actions on occasion. With the monsoon season fast approaching, a seaborne attack on Rangoon was sent in. This

GENERAL WILLIAM SLIM

Slim (1891–1970) was the principal British general in Burma for most of the war. After holding junior commands in East Africa and Syria in 1940–1, Slim came to Burma in April 1942 when the great retreat was already well under way. He took over the newly formed Fourteenth Army in October 1943 and led it with great skill until the end of the war. Unusually for a top commander, he was a modest man. He was greatly liked and respected by those he led who called him "Uncle Bill".

Above: General Slim in characteristic resolute pose.

took the city (which had been evacuated by the Japanese) shortly before Fourteenth Army arrived from the north.

The Japanese troops defending the coastal region were now cut off and trying to escape to the east. In the last fighting of the campaign before Japan finally surrendered, thousands of these men were killed at the cost of little more than a handful of Allied casualties.

Transport Aircraft and Gliders

Air supply operations on an entirely new scale were possible in WWII, in Burma and China and many other places. Mass airborne operations also began, using aircraft and gliders to bring thousands of troops suddenly into action.

World War II was the first conflict in which air transport played a significant role. It was also the first in which airborne warfare, using parachute and glider-borne troops, was employed in major battles.

TRANSPORT AIRCRAFT

The best-known transport aircraft of the early-war years was Germany's Junkers 52. Dating back to the early 1930s and with an unusual three-engined configuration, the Ju 52 had modest flight performance and could carry 18 fully armed troops. It dropped paratroops and towed gliders in Germany's early campaigns, but was less successful as a cargo carrier ferrying supplies to the surrounded Sixth Army at Stalingrad.

Below: A Ju 52 on a Russian airfield, 1941–2. Ju 52s supplied many surrounded troops that winter but failed to repeat the feat in 1942–3.

Above: A flight of Horsa gliders being towed aloft (by converted Whitley bombers) in training in 1943.

Germany also had the huge six-engined Messerschmitt 323, converted from a glider. It could carry an unequalled 21 tonnes of cargo but was very slow and vulnerable to fighter attack.

Britain depended mainly on American designs for its transport aircraft but did employ some converted heavy bombers in the role. Whitleys, Stirlings and Halifaxes all towed gliders and dropped paratroops; Avro York conversions of the Lancaster could carry 10 tonnes of cargo. The Armstrong Whitworth Albemarle, originally intended as a bomber, saw most service as a glider tug.

The USA made far more use of transport aircraft than any other nation, as well as supplying

many of these to all the Allies. The oldest design, based on the Douglas DC-3 airliner of the mid-1930s, served as the C-47 Skytrain (or Dakota in British service). Over 10,000 were made and served in all transport roles in all theatres (when configured to carry paratroops they were officially known as C-53 Skytroopers).

Serving mainly against Japan was a second twin-engined design, the Curtiss C-46 Commando. Though it was faster and could carry much

C-54 SKYMASTER

The C-54 was developed from the pre-war Douglas DC-4 airliner, first flown in 1938. The first C-54 entered service in 1942 and in all 1,170 were built. It also served with the US Navy as the R5D and with various Allied countries. A number of C-54s were used for VIP transport, including one by President Roosevelt.

ENGINES: 4 x 1,450hp Pratt & Whitney R-2000 Cyclone radials
CRUISING SPEED: 310kph (190mph)
RANGE: 6,400km (4,000 miles)
CREW: 4
CAPACITY: 50 passengers or equivalent cargo

more cargo than the C-47, only about 3,300 were made. Most served on the notoriously demanding "Over the Hump" supply route to China. Finally, the US had the four-engined C-54 Skymaster, based on the pre-war Douglas DC-4 airliner. This could carry up to 50 personnel or an equivalent cargo and mainly operated to and from bases in the USA.

GLIDERS

Even if at first glance they might seem dangerously unsuited to military uses, gliders had valuable attributes that helped them see much effective war service. They could carry significant numbers of troops into action and land each planeload in a concentrated group (paratroops might be scattered far and wide); they flew silently and could land very accurately beside, or even on top of, an objective (as was done by the Germans at Eben Emael in 1940 and by various British units on D-Day); and the troops they carried did not need to be carefully selected or given specialized parachute training.

Above: US paratroops embark in a C-53 Skytrooper. A typical load was 15–18 troops and their equipment.

Countries using significant numbers of gliders were Britain, Germany and the USA.

Germany's main glider was the DFS 230. It could carry ten men, including the pilot. (Glider pilots in all nations were usually expected to get out and fight after landing.) Larger types included the Gotha 242 and the Me 321 Gigant, parent design of the Me 323 described above. The Gigant needed three Me 110s or a specially adapted Heinkel 111 to tow it aloft, so it was not a success.

British gliders included the Airspeed Horsa, which could carry 30 men or an anti-tank gun or similar cargo, and the much larger General Aircraft Hamilcar, which could lift a light tank. The main American glider (over 12,000 built) was the CG-4A Waco, which could carry 15 men or an equivalent cargo. British and American gliders played a vital part in the invasions of Sicily and Normandy and other major airborne operations.

Iwo Jima and Okinawa

The landings on these two islands were the last major operations planned before the invasion of the main parts of the Japanese homeland. Despite their disparity in size and terrain, both saw ferocious battles before the American forces gained control.

By 1945 American planners were clear that winning the war depended on taking bases from which the invasion of Japan could be mounted and supported. Iwo Jima offered airfields within fighter range of Tokyo; Okinawa had base and harbour areas suitable for supporting the vast force needed for the final invasion operations.

IWO JIMA
No more than a dot on the map of the Pacific, closer up Iwo Jima island is pear-shaped, 8km (5 miles) long and a maximum of 3km (2 miles) wide. In 1945

KEY FACTS – OKINAWA

PLACE: Okinawa, Ryukyu Islands

DATE: 1 April–21 June 1945

OUTCOME: US forces captured a final base for the invasion of Japan after very heavy casualties on both sides.

THE CAPTURE OF OKINAWA
The Japanese forces chose to defend only the south of the island in strength.

it had three airstrips but was mostly barren volcanic rock, with a 150m (500ft) extinct volcano, Mount Suribachi, at the southern end. In 1945 it was also perhaps the most heavily fortified area in the history of warfare, with a garrison of 21,000 men under General Kuribayashi Tadamichi, prepared to defend it to the last.

Kuribayashi's defences survived the extensive preliminary air and sea bombardment – he planned to let the attackers land before opening fire and revealing his positions. The landings began on 19 February with some 30,000 US Marines of V Amphibious Corps going ashore on the first day. Soon the Marines were being hit from all directions from the complex of trenches, tunnels and other strong points, which riddled the island. The result of the fighting was never in doubt, but the battle was vicious and bloody.

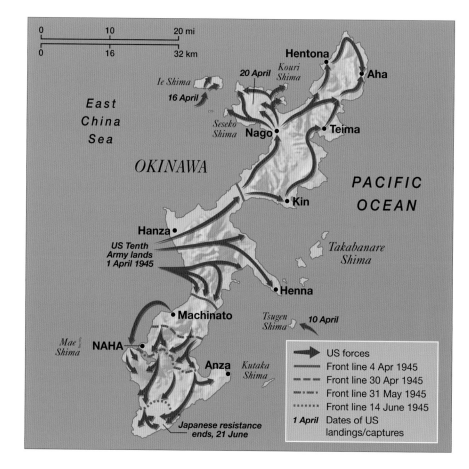

Above: A rocket-armed landing ship – LSM(R) – bombarding the Okinawa beaches, 1 April 1945.

Right: Smoke rises above Iwo Jima as US Marine amphibians approach the shore.

It took until 26 March to wipe out the defenders and cost the Marines almost 6,000 dead and over 17,000 wounded. By then, however, P-51 Mustang fighters were already operating from the island and its airfields were also providing emergency landing grounds for the B-29 force flying from the Marianas.

OKINAWA

The next target was a different and much bigger proposition. Although Okinawa is 560km (350 miles) from the nearest of the main Japanese Home Islands, it is part of "mainland" Japan (as is Iwo Jima). It would therefore be fiercely defended.

The Thirty-second Army defending the island, under the command of General Ushijima Mitsuru, was 120,000 strong. He had no intention of fighting on the beaches, for he had too much respect for American firepower.

Below: A kamikaze about to strike (and lightly damage) the battleship *Missouri* off Okinawa, 17 April 1945.

Instead he concentrated his troops in a relatively small but heavily fortified area of the south of the island, where the broken terrain would aid the defence in every way.

In all some half a million US troops and over 1,200 warships (including a significant British contingent) took part in the attack on Okinawa and various smaller islands nearby.

The naval forces suffered a blizzard of kamikaze attacks that sunk or damaged over 400 ships but never seriously disrupted Allied plans. The giant battleship *Yamato* was used as a suicide vessel, sailing from Japan with only enough fuel to reach Okinawa – it was sent to the bottom by air attacks on 7 April before it even came close.

The American landings had begun on 1 April and overran the northern three-quarters of the island by the middle of the month with relative ease. By then much harder fighting had started on the so-called Shuri Line to the south. It took two more months of vicious close-quarter battle before Japanese resistance ceased.

The US forces lost 12,500 dead on land and at sea and 35,000 wounded – frightening totals, considering the much bigger task of invading Japan itself that was in prospect.

However, the cost on the Japanese side was immense. Almost all the garrison troops were killed, as well as numerous civilians. Many more civilians committed suicide. As a final horror, among these were hundreds of children killed by their parents, who did not want them to suffer the brutalities that they had been told the Americans would commit.

Above: American troops riding up to the battlefront aboard a Sherman tank, Okinawa, 5 May 1945.

Destroyers

Destroyers were multi-purpose warships, fast and deadly hunters of surface ships and submarines with their torpedo and depth-charge armament. They served in their hundreds and saw combat after combat in every theatre of war.

Destroyers were invented to protect battlefleets from torpedo attack, and to carry out such attacks themselves. These essentially remained their main functions in WWII, though by then of course torpedo attack could come from submarines as well as surface craft.

Destroyers of WWII were generally 1,500–2,000 tons displacement and typically carried 4–6 main guns of 127mm (5in) or similar calibre and 8–10 torpedo tubes (TT), and had a top speed of about 35 knots. All also carried depth-charge equipment and a number of lighter anti-aircraft guns. More of both these types of weapon generally were added as the war progressed, along with new and improved varieties of radar and sonar equipment. Britain, the USA and Japan each began the war with over 100 destroyers in service, built many more in the course of the conflict – and each lost more than 100 in combat.

EUROPEAN DESIGNS

Among European navies Britain built relatively small destroyers in the inter-war years. Known as the A to I classes (a class of nine or so was built yearly during the 1930s), these 1,400-ton ships carried 4 x 4.7in (119mm) guns and 8 or 10 TT. However, the 4.7in gun could not be used for AA defence, so these ships were poorly equipped to withstand air attack. Britain also had the larger Tribal class with 8 x 4.7in guns and a reduced torpedo armament. Both France and Italy built some extremely large destroyers, including the Fantasque and Navigatori classes; these were large and very fast ships but paid a price in seaworthiness and reliability.

HMAS *NESTOR*

Built in Britain and commissioned in 1941, the *Nestor* served in the *Bismarck* chase and was sunk during a Malta convoy operation in June 1942. *Nestor* was one of 24 similar J, K and N class ships built for the British and Allied navies, 1937–42. Although an Australian Navy ship, *Nestor* never visited Australia.

DISPLACEMENT: 1,770 tons standard, 2,300 tons full load
CREW: 180
SPEED: 36 knots
RANGE: 5,500nm (10,200km) at 15 knots (27.8kph)
LENGTH: 111.4m (356.5ft)
BEAM: 10.9m (35.75ft)
ARMAMENT: 6 x 4.7in (119mm) guns + 5 x 21in (533 mm) TT; 1 x 4in (102mm) AA gun + numerous light AA weapons

Left: The Kagero-class *Yukikaze*, pictured in January 1940. *Yukikaze* was the only one of the 19 ships in the class to survive the war.

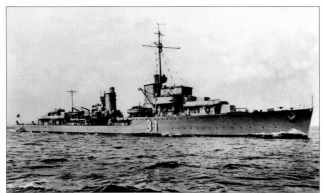

Above: USS *Fletcher* in July 1942. The 177-ship Fletcher class was the most numerous destroyer type ever.

Above: The German destroyer *Z14* or *Friedrich Ihn* pictured in 1942, one of 12 ships of the 1934A class.

Germany's pre-war and war-time ships were a mix of larger vessels classed as "destroyers" and smaller "torpedo boats". Destroyer or *Zerstörer* types were typically 2,400 tons, with 5 x 12.7cm (5in) guns. British war-time destroyers, in lettered classes up to W, were generally slightly larger than their predecessors, with better AA capability. Some had 4in (102mm) AA guns as main armament instead of the 4.7in weapons.

PACIFIC NAVIES

Japan and the USA favoured slightly larger ships than the British for their pre-war classes and had the advantage of having suitable dual-purpose (DP) surface/AA gun mounts with which to arm them. Japan also had the 610mm (24in) "Long Lance" torpedo, far superior to anything in Allied service. In addition the Japanese fitted their ships with torpedo reloading equipment that could be used in action, allowing multiple attacks to be carried out.

Right: The USS *Wilson*, one of ten Benham-class destroyers of the US Navy, seen in January 1941.

Japan's Fubuki class (24 completed by 1932) were the most powerful destroyers in the world when built (1,800 tons, 6 x 127mm, 9 TT); they proved too flimsy so were extensively rebuilt early in their service. Another notable class was the 2,000-ton Kagero type with similar armament to the Fubukis.

The US Navy had the best DP main gun: the 5in/38-calibre Mark 12, fitted in most US destroyers from the Farragut class (1934, 1,400 tons, 5 x 5in, 8 TT) onward. The Porter class of 1936–7 was larger (1,850 tons, 8 x 5in), in an attempt to match the big Japanese designs, but these destroyers had stability and seaworthiness problems.

However, US Navy classes built during the war years were the best destroyers of the time. The Fletcher class (2,100 tons, 5 x 5in, 10 TT, 35 knots) had an excellent balance of firepower, stability and speed, and the capacity to fit additional AA guns and other equipment shown to be necessary by wartime experience. The later Allen M. Sumner and Gearing classes were slightly bigger but generally similar, with the exception that the Fletchers' five single 5in mounts were replaced by three twin turrets.

Weapons of Mass Destruction

Although thousands had died from the effects of poison gas in WWI and cities were razed to the ground by conventional bombing attacks in WWII, the atomic, chemical and biological weapons developed by 1945 were a new and more terrible threat.

By the late 1930s scientists were familiar with the idea that chain reactions might be created in certain elements to give off vast amounts of energy that could conceivably be used in some type of bomb.

Scientists in Japan and the USSR were among those who realized this possibility, but neither of those countries pursued the idea at that stage – and American research was not then extensive.

Britain, however, took things further. Rudolph Peierls and Otto Frisch, Jewish refugees from the Nazis, made the breakthrough; they calculated that a relatively small quantity of the rare uranium 235 isotope would be needed for a bomb. (Ironically, the two were only working in this field because they were not yet fully trusted to join native British scientists in "more vital" electronics and radar research.) Other scientists, in Britain and the USA, also

Above: General Leslie Groves (right) and J. R. Oppenheimer, respectively military head and chief scientist of the Manhattan Project.

worked out a second method of "bomb-making". This involved the creation of a new element – plutonium – from the common uranium 238 isotope.

THE FIRST A-BOMBS

British developments were shared with the Americans, who began more serious work in late 1941. This soon developed into the huge Manhattan Project, to which the British scientists were transferred. By early 1945 both U235 and plutonium were being produced in sufficient quantities and two designs of bomb were being finalized.

On 16 July 1945 a plutonium device was exploded in a test at Alamogordo, New Mexico. Its yield was calculated as equivalent to at least 15,000 tonnes of TNT; the explosion was visible and audible up to 275km (170 miles) away. The bomb dropped on Nagasaki, known as "Fat Man", was a second plutonium weapon. The Hiroshima device, "Little Boy", was based on the U235 isotope.

Although the greatest motivation in starting the research was to forestall German development of similar weapons, US leaders soon realized that the unprecedented power of atomic weapons could transform international affairs. Ideas that Japan be given a demonstration of the bomb's power before it was used were never taken seriously. With Allied lives at stake, there was never any question that the bomb would be used in action as soon as possible.

GASES AND POISONS

WWII also saw developments in chemical and biological warfare. Fortunately such weapons were little used, mainly from fear that any use would bring equivalent retaliation and therefore gain nothing. However, it was a great relief to Allied

Below: The two atomic bombs used against Japan in 1945. "Little Boy" dropped on Hiroshima (right), and "Fat Man", the Nagasaki weapon (left).

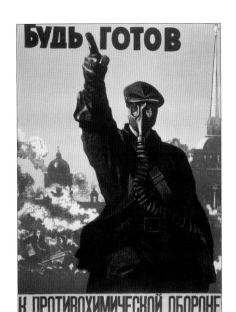

Above: "Be prepared to ward off chemical weapons" – a Soviet poster from WWII.

HEAVY WATER

One method of making plutonium used so-called "heavy water" (a compound featuring the hydrogen isotope deuterium), for which the best source under German control was a hydroelectric plant at Rjukan in Norway. British and Norwegian special forces sabotaged the plant in 1943 and in 1944 prevented its production reaching Germany. By then, unknown to the Allies, Germany's atomic research had been effectively abandoned because the amount of $U235$ needed for a bomb had been miscalculated.

leaders that Germany's V-weapons had only conventional explosive warheads.

All nations took precautions against the use of poison gas, as they had in WWI. Gas masks were issued to troops and civilians; generally stocks of WWI gases like mustard gas and phosgene were ready for retaliatory use. Italy employed mustard gas in its conquest of Abyssinia, but such weapons were not used otherwise in WWII. Germany alone developed nerve gases – tabun and sarin – but did not deploy them. This research fell into Soviet hands in 1945.

Several countries experimented with biological agents and produced weapons to use them, though Hitler forbade such research in Germany. Both Britain and the USA had anthrax and botulin weapons ready by the end of the war, but only Japan used weapons of this type. The Japanese researchers (known as Unit 731 and commanded by General Ishii Shiro) killed many hundreds of Chinese prisoners of war in experiments in this field. They made successful attempts to spread diseases like cholera, typhoid and plague against the Chinese forces, notably in 1942 in Kiangsi province.

Above: Governments took the threat of chemical attack seriously as this 1943 British exercise shows.

Ishii and his team were captured by the Americans in 1945 and given immunity from prosecution in return for information about their research.

The Japanese Surrender

Japan had already suffered greatly by August 1945, but then came a succession of hammer blows: the atomic attacks on Hiroshima and Nagasaki and the Soviet declaration of war and subsequent crushing land offensive in Manchuria.

Ever since the Casablanca Conference in January 1943 the official Anglo-American policy was to seek the unconditional surrender of Japan. At that mid-war period, facing the prospect of a longer struggle to close in on Japan than in fact transpired, both the USA and Britain were keen to persuade the Soviets to join the war in the Far East. With offers of territorial and other concessions, Stalin duly promised at the Teheran Conference in late 1943 to declare war on Japan shortly after the war in Europe had been won.

When the Allied leaders met again at Potsdam in July 1945 much had changed: Hitler was dead in the ruins of nearby Berlin; Japan was far nearer to defeat in a shorter time than had once seemed possible; and on

Below: The centre of Hiroshima. Only a few stoutly constructed buildings still stand after the attack.

Above: The distinctive atomic mushroom cloud forms above the devastated city of Nagasaki.

the day before the conference started, the world's first nuclear bomb was tested in New Mexico. News of its power was rushed to the new US President Harry Truman and the British leaders at Potsdam.

The Japanese government had already begun making peace feelers through its embassy in Moscow but it was clear from these that they wanted guarantees that Emperor Hirohito's position would not be at risk. The Allies could not accept this, feeling that to do so might be taken as a sign of weakness, which would encourage prolonged resistance.

Instead they issued the "Potsdam Declaration", which mentioned nothing about the Emperor but merely repeated the demand for unconditional surrender, threatening unspecfied "utter destruction" if the demand was not met. In the meantime Stalin was officially told (though he already knew through his spies) that the USA had a new and powerful weapon ready for use against Japan and confirmed that he would soon be ready to attack Japan.

When the Japanese Prime Minister, Admiral Suzuki Kantaro, replied publicly to the declaration he said he would not comment on it for the moment; but he was misunderstood as saying that Japan would ignore it. With that Truman gave the order for the bomb to be used.

ATOMIC ATTACKS
The first atomic bomb was dropped on Hiroshima on the morning of 6 August 1945 by a B-29 bomber flown from the island of Tinian. The bomb killed about 80,000 people more or less instantly; an estimated 50,000 more died from its short- and long-term radiation effects.

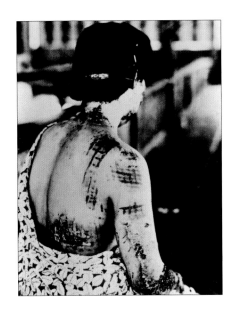

Above: Many of the survivors, like this woman from Hiroshima, suffered terrible burns in the atomic attacks.

Nagasaki suffered a similar fate, but with slightly fewer casualties, on 9 August.

There is no doubt that Truman's decision to use the bomb was made in large part in the hope that it would spare lives by making Japan surrender. However, the historical record is also clear that part of the motive was to intimidate the Soviets and lay down a marker for the post-war world.

The Soviets issued their promised declaration of war on the 8th and their forces crashed across the border into Japanese-held Manchuria the next day. Japan's supposedly elite Kwantung Army, outnumbered and outmatched in every category of weaponry, simply fell apart. Manchuria and northern Korea were overrun within days.

THE SURRENDER

Emperor Hirohito now intervened to tell the diehard militarists in his government that the war must be ended. To the last there was the same dithering and failure to face facts at the heart of Japan's government as had helped get Japan into the mess in the first place. This was epitomized in Hirohito's broadcast to his people in which he blandly stated that "the war situation has developed not necessarily to Japan's advantage".

Even so that same day, 15 August, was celebrated by the Allies as VJ-Day. Japan's formal surrender was signed on the US battleship *Missouri* in Tokyo Bay on 2 September.

Below: General MacArthur, backed by a row of Allied representatives, at the surrender ceremony.

INVASION OF JAPAN

The planned Allied invasion of Japan would have been the largest landing operation in history. Two phases were intended: Operation Olympic against the island of Kyushu in November 1945, and, using bases gained there, Operation Coronet, the final decisive attack on the Kanto Plain on Honshu, near Tokyo, scheduled for March 1946. Casualty estimates varied widely, but typical figures were 400,000 Allied dead (including significant British Empire forces, not just Americans) and more than ten times that number of Japanese.

New Nations Refugees at Amritsar in the last days of pre-independence India.

Dealing with the Defeated Leading Nazis in the dock during the Nuremberg trials.

The Cold War American transport aircraft unloading supplies in Berlin during the Airlift in 1948.

NATO members 1955
USA and Canada also NATO members

Warsaw Pact members 1955

| 0 | 100 | 200 | 300 | 400 | 500 mi |

| 0 | 200 | 400 | 600 | 800 km |

ICELAND

ATLANTIC OCEAN

NORWAY

SWEDEN

FINLAND

IRELAND

UNITED KINGDOM

North Sea

DENMARK

Baltic Sea

USSR

NETHERLANDS

BELGIUM

EAST GERMANY

POLAND

LUXEMBOURG

WEST GERMANY

CZECHOSLOVAKIA

FRANCE

SWITZERLAND

AUSTRIA

HUNGARY

ROMANIA

PORTUGAL

SPAIN

Corsica

ITALY

YUGOSLAVIA

ALBANIA

BULGARIA

Black Sea

Sardinia

GREECE

TURKEY

IRAN

Mediterranean Sea

Sicily

SYRIA

MOROCCO

CYPRUS

IRAQ

ALGERIA

TUNISIA

ISRAEL

JORDAN

LIBYA

THE LEGACY OF WAR

No one claimed in 1945 that World War II would become the "war to end all wars", but it did establish that any future war would have different causes. The expansionist militarism that had characterized both Germany and Japan since the early years of the 20th century was discredited for good in the eyes of most German and Japanese people, as was the murderous racism that accompanied it.

For a time after 1945 it also seemed that the US atomic monopoly (in the event lasting only until 1949) would deter attacks on the USA and its friends. However, the mutual suspicions between the USSR and the Western Allies (some of them justified, some not) brought the increasing tension and competition of the Cold War. Although the Cold War itself did not outlast the century, its legacy remains important in the new millennium.

Many of the more controversial methods employed in modern conflicts can trace their history directly to WWII. The resistance movements and similar organizations of the war used tactics familiar 60 years on to freedom fighters and terrorists and, for all their precision weapons, 21st-century air forces still bomb civilians as regularly and with as little scruple as their forebears did.

Dealing with the Defeated
Temporary housing being built amid the devastation of Hiroshima.

The Cold War President Truman signs the NATO treaty, committing the USA to the defence of Europe.

Europe Divided German civilians trying desperately to gather coal for fuel on a mine spoil heap.

Casualties and Destruction

World War II was the most brutal conflict in human history, with a death toll several times that of any previous war. Death came to civilians far from the front line, as it did to soldiers, with a dreadful thoroughness never before contemplated.

The human cost of WWII was truly enormous but is still impossible to measure exactly. Even in the early 21st century, when most records have been opened to historical study and most of the bitterness associated with the war has long subsided, no one has been able to calculate exact figures, nor ever will be. The lowest figure that is at all plausible is some 40 million dead, but other estimates go as high as 55 million. Also on top of that huge figure were probably three times as many wounded, of whom many would have to live with pain and disability for the remainder of their lives. And as well as these were the many more still, who were emotionally scarred by their experiences or the loss of loved ones. The weight of suffering was immense.

HARROWING TOLL

There is no question that the losses of the Soviet Union were the greatest of all and that the

Above: A memorial to the children of the Czech village of Lidice, destroyed by the Nazis in 1942.

Eastern Front was the theatre in which the fighting was the largest in scale and the least restrained by moral limitations. The Soviet armed forces lost at least 10 million killed, including a large proportion who died of ill-treatment as prisoners in German hands and many who, once liberated from German captivity or returned to Soviet control by the Western Allies, were sent straight to the GUlag. Soviet civilian deaths amounted to 10 million or more, the largest proportion of these being murdered by Stalin's regime.

Dominating the list of European civilian casualties are the 6 million Jews murdered by the Nazis, who came all too close to their objective of making Europe Jew-free – only around 300,000 Jews from areas that had been Nazi-controlled survived the war.

The country second on the list after the USSR is China, another for which any exact reckoning is impossible. Common estimates are in the range of 10–15 million, of whom 2–3 million were fighters in one or other of the armies.

The major Axis powers both fared badly, though in each case military casualties predominated over civilian. Perhaps 4.5 million German soldiers died and another 2 million civilians. Japan's total casualties were about 2 million. Poland and Yugoslavia both suffered particularly severely. Poland lost some 4–4.5 million, more than half of them Jews, and Yugoslavia over 1.5 million in the country's vicious resistance and civil-war struggle.

Of the major Allied powers, Britain lost some 350,000 dead, considerably fewer than in WWI, and the countries of the British Empire an additional 120,000. In proportion to its population the losses of the USA were by far the lowest of any

Left: A memorial at the former Dachau concentration camp near Munich, one of the first to be set up.

1933–1945

major combatant: approximately 275,000, of whom all but a handful were military personnel.

In addition to the millions killed or injured were the millions more who ended the war as refugees. Even if they did have homes to return to, many had no wish to do this because wartime events or subsequent political changes meant they would be in danger or unwelcome. There may have been as many as 30 million "displaced persons" (as they were known) in Europe at the end of the war and millions more in China.

PHYSICAL DESTRUCTION

As well as the human cost of the war the scale of physical damage was stunning. Even in Britain, which was comparatively lightly bombed and saw no land fighting, hundreds of thousands of homes suffered war damage. In Germany and Japan whole cities were virtually demolished by Allied bombing. In the western

Above: The British War Graves Commission cemetery at El Alamein, seen shortly after the war.

USSR roughly two-thirds of the homes, factories and other resources were destroyed in the fighting or by the scorched-earth policies applied in retreat, first by the Soviets themselves and later by the Germans.

Left: A British soldier returns to his family, in their new "prefab" home, after a four-year absence.

Right: Displaced persons crowding a train in Berlin in June 1945.

Dealing with the Defeated

*Unlike in WWI, when the Allies only occupied part of Germany, in 1945 both
Germany and Japan were occupied and administered by the Allied powers.
International tribunals also tried and punished war criminals in both countries.*

At the Casablanca Conference in January 1943 President Roosevelt had declared (and Prime Minister Churchill agreed) that the Allies would seek the unconditional surrender of Germany and Japan. Later in 1943 Britain, the USA and the USSR established the European Advisory Commission to draw up surrender terms for Germany and to work out occupation arrangements for Germany (and Austria, which the Allies decided should be separated from Germany again) and the minor Axis powers. These plans were eventually agreed by the heads of government at the Yalta and Potsdam conferences in 1945.

Allied Control Commissions were set up to run the occupied Axis countries. In the case of Germany this meant that the country was to be divided into

Above: July 1945, a typical Berlin street scene. Locals trudge past the rubble with their few belongings.

four occupation zones – one each for the USA, UK, USSR and France – and Berlin would also be similarly divided.

Various plans for Japan were considered, including versions with the country divided into Allied zones. Instead, when the time came, General MacArthur was appointed Allied Supreme Commander for the occupation. All of Japan's main islands came under American control, with MacArthur acting almost as a ruling monarch. The Soviets occupied (and would retain as Soviet territory) the Kurile Islands and the southern half of Sakhalin, while the occupation forces of the main Japanese islands included a British Empire contingent.

PUNISHING THE GUILTY

In 1943 the Allies had also announced their intention to punish Nazi war criminals. In the Potsdam declaration in 1945 the process was extended to include Japanese leaders. This resulted in two sets of major international war crimes trials, at Nuremberg and Tokyo, in which some of the most senior figures were tried. In the end 22 Germans and 25 Japanese were tried (indictments in both trials named others who died, were judged unfit to plead or committed suicide). There were also numerous trials conducted by the individual Allied powers of those accused of crimes affecting only one Allied nation.

Left: Temporary housing for civilians in the area blasted by the Hiroshima atomic bomb.

The charges at the major trials included two new concepts in international law, crimes against peace and crimes against humanity, as well as the better-established concept of war crimes in breach of previous internationally agreed standards regarding such things as treatment of prisoners of war. There were three outright acquittals at Nuremberg but none at Tokyo, though some defendants were found not guilty on some of the charges brought against them. Punishments included 19 death sentences and various long terms of imprisonment.

VICTORS' JUSTICE?

Some aspects of the trials were unsatisfactory. Certainly there was an element of victors' justice – for example, Stalin was just as guilty for his invasion of Poland in 1939 and the subse-

Above: General Tojo, Japan's Prime Minister for most of the war, was executed after the Tokyo trials.

quent murders of thousands of Poles as any of the Nazis. And a truly impartial court might well have considered the Anglo-American strategic bombing campaigns as war crimes.

In the Japanese case the defendants were selected and the prosecution evidence slanted so that Emperor Hirohito

and his family were left untainted, because this suited the Allied occupation policies.

Although there has been much discussion since about the legality and fairness of the proceedings, they definitely established the principle that national leaders can be brought to account before the international community and that soldiers everywhere have a duty to disobey unlawful orders. And the evidence at Nuremberg proved beyond all doubt that the Nazis did murder some 6 million Jews among much else, however much some have since tried to deny this. Whatever flaws there may have been, these were valuable precedents.

Below: Top Nazis in the dock at the Nuremberg trials. Hermann Göring (front left) was the most senior figure to stand trial.

Europe Divided

In a famous speech in 1946 Winston Churchill described how Europe had been divided by an Iron Curtain. The suspicions that had hindered the wartime alliance between Britain, the USA and USSR were hardening into something more serious.

Although the war ended in 1945 with Western and Soviet troops in control of respective halves of Europe, very few would have expected this to be a long-lasting state of affairs. Indeed the leading Allied powers, including the Soviets, had issued the so-called Declaration on Liberated Europe at the Yalta Conference in early 1945. This stated that they intended to see democratic governments and free elections in countries they had liberated from German control.

WESTERN EUROPE

There was never any question that these pledges would be fulfilled in Western Europe. The Allied powers had backed governments-in-exile of all

Above: Germans searching through the spoil from a coal mine in a desperate hunt for scarce fuel.

Below: Czech politician Jan Masaryk died in suspicious circumstances after the communist takeover in 1948.

these countries during the war. The removal of the Germans in 1944–5 was followed shortly after by elections. Local communist parties, which had been prominent in the resistance movements to the German occupation, participated fully in these elections and, notably in France and Italy, won a significant number of votes.

It was clear that conditions in Eastern Europe would be different. Even as the Yalta Conference was taking place, the Soviets were already working to ensure that communists would take control in Bulgaria and in March 1945 they set up a communist government in Romania.

THE UNITED NATIONS

The history of the United Nations (UN) dates to the Four Power Declaration (by the USA, UK, USSR and China) in October 1943 that they intended to establish an international organization after the war to maintain "international peace and security". After further discussions the UN was established by the San Francisco Conference in 1945, attended by delegates from 50 countries. Even at that early stage there were problems. Poland was not represented because the USA and USSR did not agree about the composition of a legitimate Polish government.

Right: A US atomic test at Bikini Atoll, 1946. Old warships are placed nearby to assess the blast's power.

Britain (and later the USA) had supported a Polish government-in-exile in London from the early stages of the war but, as the Red Army advanced in 1944, the Soviets put forward an alternative communist group, which they recognized at the end of the year as the provisional government. Some of the "London Poles" did return home in 1945, but they had been removed from government and all opposition silenced by early 1947. During 1947–8 there were further communist takeovers in Hungary and Czechoslovakia.

OPPOSING COMMUNISM

Concerns over these developments and disagreements about how to deal with Germany soon provoked changes on the Anglo-American side. At the start of 1946 President Truman spoke about the need to "get tough with Russia" and other senior figures urged that the USA adopt a policy of "firm and vigilant containment of Russian expansionist tendencies". Winston Churchill's "Iron Curtain" speech followed shortly after and, though he was by then an ex-prime minister, the British government agreed with him.

Since the end of WWII Britain had been helping the Greek government to combat a communist-led civil war, but in early 1947 Britain, in dire economic trouble, told the USA that it could no longer afford to do so.

Right: President Truman addressing the first meeting of the United Nations in June 1945.

Stalin had kept his wartime promise not to interfere in Greece and such external aid as the Greek communists were receiving came from Yugoslavia, which was not at all the same thing as saying it came from Moscow. But this was not clearly understood then in the West.

Instead the USA agreed to take over Britain's role in Greece and also to help Turkey. This policy was summed up in March 1947 by President Truman's promise to "support free peoples" resisting subversion. This was the start of what became known as the Truman Doctrine.

Europe's economies as a whole were still suffering from the destruction of the war years and the particularly severe winter of 1946–7 had not helped. In 1946 Britain and the USA had agreed to stop taking reparations from Germany. By early 1947 they were beginning to consider supporting a German economic recovery as part of a concerted strategy to revive European prosperity. This would also be beneficial to the Americans because of the markets it would create.

Western and Soviet policies were rapidly moving farther apart than ever.

New Nations

Japanese rule in Asia had been brutal and racist but it made a restoration of the European colonies impossible. In the Middle East substantial Jewish immigration after the war also meant that new conditions prevailed.

The wartime slogan of Japan's Greater East Asia Co-Prosperity Sphere had been "Asia for the Asiatics" and, for all the brutality of their subsequent rule, the Japanese triumphs of 1941–2 had largely discredited the former colonial masters. The Anglo-American Atlantic Charter of 1941 had, at American insistence, openly supported the right of people everywhere to choose their own governments – and American policy throughout the war was avowedly anti-colonial. Both of these factors assisted the developing forces of nationalism.

COLONIAL STRUGGLES

In French Indo-China (now Laos, Cambodia and Vietnam), for example, the American Office of Strategic Services sup-

Above: A French soldier in Indo-China in 1951 during the fruitless attempt to maintain French rule.

Below: Refugees in Amritsar, among the millions forced to move because of their religion when India and Pakistan became independent.

ported a Vietnamese nationalist coalition led by the communist Ho Chi Minh during the war. Ho's Viet Minh forces fought against the Japanese and, at the end of the war, proclaimed the existence of a Democratic Republic of Vietnam. France subsequently tried (unsuccessfully) to reimpose colonial rule. Thus began the long conflict, which would only end with the North Vietnamese victory over South Vietnam in 1975.

In the Dutch East Indies the pre-war colonial power also tried and failed to reimpose its control after the war. Here the Japanese had worked with the nationalist leader Achmad Sukarno during the war and in 1944 had promised to grant independence. With help from the Japanese, Sukarno went on to proclaim Indonesia as an independent country in August 1945. This was finally recognized by the Dutch in 1949 after a bitter, armed struggle.

Unlike the Dutch and the French, Britain had substantial forces available to re-occupy Malaya and the other territories that remained under Japanese control in August 1945. However, whether they had been Japanese-occupied at any stage or not, almost all British colonies in Asia quickly gained their independence. In 1939, despite the well-established system of partial self-government, Britain's viceroy in India had declared the country to be

Right: British Indian Army troops fighting in Java in late 1945 against Indonesian nationalists. British forces initially took control in Indonesia after Japan's surrender.

at war with Germany without consulting a single Indian. By 1945 there was no doubt that those days were past, and in August 1947 India and Pakistan became independent.

ISRAEL AND THE ARABS

European rule had also had its day in the Middle East. The pre-war French mandate in Syria was never restored and the country became independent in effect in 1944. Britain's pre-war mandate in Palestine had been troubled by violence between Jews and Arabs and the British authorities; various plans had been proposed for separate and joint Jewish and Arab states.

British control continued as a United Nations' mandate after the war amid growing violence. Jewish immigration and the experience of the Holocaust lent more force to Zionism than previously, while UN attempts to broker a deal between Arabs and Jews came to nothing. In 1948, when the British mandate expired, the state of Israel was proclaimed; Israel had been successfully established by 1949. Perhaps around 700,000 Palestinian Arabs were displaced from their homes, creating a legacy of bitterness that persists into the 21st century.

COMMUNIST CHINA

The biggest of the post-war changes was in China. For all the Allied aid that had been sent to the Nationalists, the Communists had done better out of the war. Their armed forces were better organized and in 1945 they already had an effective presence in most of the supposedly Japanese-controlled areas of the country.

Over the following years, the communists gradually gained the upper hand throughout the country and in 1949 proclaimed the People's Republic of China; the defeated Nationalists withdrew to Taiwan. In the West this looked like an important victory for world communism, as directed from Moscow. In reality the Chinese communists were masters in their own house, but for the moment their success only contributed to the development of Cold War tensions.

Right: David Ben-Gurion, first Prime Minister of Israel.

The Cold War

By 1949 the focus of international affairs was no longer on dealing with the effects of WWII. Instead the USA, USSR and many other nations were preoccupied by a new and, in an era of nuclear weapons, possibly more dangerous conflict, the Cold War.

By the middle of 1947 the relationship between the USA and USSR was already at a low ebb. The two had never entirely trusted each other as wartime allies and things had since got much worse. For the moment the USA still had a monopoly of atomic weapons and could be fairly sure that it and its friends were immune from any direct military attack.

However, Europe's economies were still failing to make any substantial recovery from the disastrous conditions of 1945. In this atmosphere, American leaders feared that communist ideas might make headway. A central part of the problem was that Allied policies had in fact prevented the reconstruction of Germany. Instead, led by the Soviets, the Allies had stripped German industry of much plant and other resources.

Above: Konrad Adenauer, first Chancellor of the new nation of West Germany, is congratulated by a supporter after his election victory in August 1949.

THE MARSHALL PLAN

This all changed with the announcement of the Marshall Plan. By its terms, American aid was to be offered to all the nations of Europe, including the USSR. President Truman and his advisers attached conditions to the aid – among other things countries receiving it would have to submit to American inspections of how the money was being spent – which they correctly calculated would make it unacceptable to the Soviets. They duly rejected it and ensured that their satellites in eastern Europe did so too.

Marshall Plan aid began to flow in 1948 and over $12 billion would be provided to a range of countries until 1951. During these years the European economies grew substantially and levels of prosperity surpassed those of the late 1930s. International trade also revived, to the benefit not just of the Europeans but also of the American economy. And finally, because they had been forced to work together to decide how the aid should be split up, the European nations began the process of economic co-operation that would lead to the establishment in 1957 of the European Economic Community and in time to the present-day European Union.

NEW GERMANIES

In March 1948 the USA, Britain and France agreed to merge their occupation zones of Germany into a single entity. In June they introduced a new

Left: Unloading transport aircraft at Tempelhof airfield, June 1948, during the Berlin Blockade.

GEORGE MARSHALL

General Marshall (1880–1959) was Chief of Staff of the US Army from 1939 to the end of the war. He is regarded as the main organizer of the US victory in WWII. First, in 1939–41, he substantially expanded and updated his service and laid the ground for the greater expansion that followed. Throughout the war he worked hard to maintain the principles of "Germany first" and that Germany should be defeated by a direct cross-Channel invasion. After the war he served as Secretary of State, helping devise and implement the Foreign Assistance Act, which came to be known as the Marshall Plan.

Above: General George Marshall in US Army uniform late in the war.

currency for the whole area to replace the Hitler-era Reichsmarks that were still in use.

Since the main aim of Soviet policy in Europe after the war was to ensure that Germany could never again dominate the continent, any measures that presaged a German economic revival were seen as threatening. The Soviet response to the Allied currency reforms and other measures was to close all road and rail routes to the Allied sectors in Berlin. This Berlin Blockade is commonly seen as the first clear confrontation of the Cold War.

The Soviets presumably hoped that the Allies would back down over their plans for Germany. They had not reckoned on the ability of the Allies to supply the city by air. Hundreds of American and British aircraft made daily transport flights into the city carrying everything that the Berliners needed. Thus supplied, the city held out until the Soviets relented in 1949.

By then the Cold War had truly begun. In 1949 West and East Germany were established as new countries, with the first elections in the West in August. In April, 12 countries, including Britain, the USA and Canada, signed the North Atlantic Treaty, agreeing that an attack on any one would be regarded as an attack on all. This NATO alliance was an open commitment by the USA not to return to the policies of isolationism that had prevailed before WWII. On the other side there was a new threat. In August 1949 the Soviets tested their first atomic weapon and would establish their own military alliance of Soviet-bloc countries, the Warsaw Pact, in the 1950s.

The Cold War competition between NATO and the Warsaw Pact and the accompanying nuclear balance of terror would dominate international affairs for the next forty years. The great issues that had brought about WWII were now relegated to history.

This edition is published by Lorenz Books, an imprint of Anness Publishing Ltd, info@anness.com

www.lorenzbooks.com;
www.annesspublishing.com

© Anness Publishing Ltd 2019

A CIP catalogue record for this book is available from the British Library.

Anness Publishing
Publisher: Joanna Lorenz
Editorial Director: Helen Sudell

Produced for Lorenz Books by
Toucan Books

Toucan Books
Managing Director: Ellen Dupont
Editor: Marion Dent
Project Manager: Hannah Bowen
Designer: Elizabeth Healey
Cartographer: Julian Baker
Picture Researcher: Debra Weatherley
Proofreader: Theresa Bebbington
Indexer: Michael Dent

PUBLISHER'S NOTE
Although the information in this book is believed to be accurate and true at the time of going to press, neither the author nor the publisher can accept any legal responsibility or liability for any errors or omissions that may have been made.

PICTURE CREDITS

akg-images front cover tl and tr, back cover bl and br, 5bmr, 6t, 8r, 9tl, 10tm, 12t, 13t, 15b, 19b, 24b, 26tl, 27l, m and r, 31t, 33t, 36t, 38b, 41r, 46b, 57b, 61t, 62t, 65t, 69bl, 71l, 78t, 96t and b, 97tl and b, 102r, 105t, 106t, 107bl, 110bl, 111tl, 119b, 123tr, 136b, 138t and b, 152tl, 164t, 165t, 166b, 167t, 171t, 173b, 181b, 182t and b, 184l, 226b, 239tl, 241r, 250b, 253, /ullstein bild back cover tmr, 15t, 31b, 32t, 37t, 45b, 48t, 66r, 67b, 71r, 77t, 81b, 94tm, 95bl, 100, 103b, 104b, 108b, 109t, 119t, 132br, 133t, 135bl and br, 141tl, 151b, 156t, 165b, 174b, 176b, 179tl, 184r, 189b, 194t, 204r, 219t, **Aviation-images.com** 42r, 121tl, 222b, /H.Cowin 39l and r, 49t, 140r, 141r, 147t, 190b, 191m, /P. Jarrett 43b, 48b, 49b, 160b, 161b, 175tl, /Royal Aeronautical Society 160t, /TRHP 175tr, /Mark Wagner 121tr, /R. Winslade 153l, 175b, **CORBIS** 4bml, 9bml, 24t, 28r, 42l, 74, 75b, 89b, 91bl, 92t, 101b, 123b, 129tr, 152tm, 156b, 159b, 170b, 171b, 182, 187l, 189t, 200tm, 201m, 206t and b, 207t, 209t, 211t, 213tr, 215l, 216b, 220r, 224b, 231b, 233br, 241m, 247b, 251t, /Bettmann back flap, 3tr, 4br, 8mr, 9tmr, 9bl, 12b, 14t, 18t, 20t, 20b, 21bl, 25t, 30l, 56t, 57t, 59b, 60t and bl, 63t, 70tm and tr, 73br, 79t, 82t and b, 85b, 90, 91br, 94tr, 95m, 110r, 117t, 130b, 131b, 142t and b, 149t, 152tr, 154t, 155tr, 158b, 162b, 167br, 179b, 185b, 191t, 194b, 201l, 203br, 210b, 214t and bl, 217b, 218t, 222t, 224t, 225b, 238b, 240tl and tm, 245t and b, 248t and b, 249b, 251b, /Stefano Bianchetti 8l, 11m, 16t, /The Dmitri Baltermants Collection

107t, 183tl, /DPA 99b, 102bl, 240tr, 246t, /EPA 239tl, /Christel Gerstenberg 4bl, 137b, /Godong/ Franois Galland 5bl, 242b, /Hulton-Deutsch Collection front flap, 5bml and br, 17t, 18bl, 22t and b, 93tl, 98t, 117bl, 153r, 177t, 197t and b, 198, 209b, 227b, 247t, 250t, /Minnesota Historical Society 61b, /Museum of Flight 202b, /Robert Harding World Imagery/Richard Nebesky 9br, 242t, /Scheufler Collection 10tl, 21br, /Swim Ink 2, LLC 46t, /Underwood & Underwood 92b, **Getty Images** /AFP 162t, /Hulton Archive front cover tml and tmr, back cover tl, tr, bml and bmr, 1m and r, 3tl, 6l and r, 8ml, 9tml, 10tr, 11l, 13b, 14b, 16b, 17b, 18br, 23t and b, 25bl, 27m, 28l, 29t and b, 32b, 40t and b, 41l, 43t, 44b, 47t, bl and br, 50t and br, 51b, 56b, 60br, 63br, 64br, 65b, 68t and b, 69t and bl, 72b, 73t and bl, 75tl and tr, 80t, 81t, 94tl, 97tl, 99t, 103t, 107br, 113bl, 14bl, 115tr, 118, 120bl, 120br, 130t, 131t, 139r, 140l, 143tr, 146b, 148t, 153m, 158t, 159t, 163t and b, 164b, 166t, 170t, 173m, 174b, 176t, 185t, 192r, 194t, 195t, 196b, 199b, 201r, 205b, 208b, 213tl, 223b, 225tl, 229t, 230t, 231t, 232r, 236t, 236br, 237b, 238t, 243t and bl, 244t, 256, /Roger Viollet 11r, 21t, 111tr, /Silver Screen Collection/Hulton Archive 111b, 254, /Time & Life Pictures 1l, 2, 7b, 9tr, 26tm and tr, 45t, 66bl, 72t, 78b, 79b, 83b, 85tl and tr, 88t, 116, 117br, 123tl, 135t, 139t, 143b, 146t, 150, 161t, 188b, 200tl, 202t, 203bl, 205t, 207b, 208t, 210t, 212, 218b, 221b, 223t, 225tr, 226t, 227t, 233bl, 235b, 241l, 243br, 244b, 246b, 255, **Photographs reproduced by permission of the Trustees of The Imperial War**

Museum, London.
Front cover b MOI FLM 1531, back cover tml TR 1402, 3tm A 30523, 30r E 1416, 35l A 6165, 44t C 1653, 52b FL 14721, 53b A 28203, 54t E 100, 55t CM164, 55b E 1376, 58b A 24047, 62b HU 1761, 67t H 23836, 71m A 30523, 76br A 30080, 80b HU 2762, 87t A 2298, 93tr A 30523, 101t SE 358, 105r B 15007, 113t E 14775, 114t BU 823, 115b H 36806, 122 MH 29100, 126bl FL 1204, 127t C 4050, 128t A 2003, 132l BU 2756, 133tr B 15229, 136t TR 1800, 137t H 25860, 143tl HU 4052, 145tl TR 450, 145tr H 34424, 147b TR 37, 151t TR 1402, 155tl B 5090, 157br H 37859, 172 CL 1756, 173t C 4919, 177b E (MOS) 1403, 179tr EA 44531, 181t BU 127, 183tr E 19296, 190t CL 2946, 217t A 21447, 219b IND 4723, 228r SE 3073, 230b COL 353, 234t A 9060, 236bl MH 6810, 249t SE 5661, **Mirrorpix** 229b, **Photo12.com** /Keystone Pressedienst 63bl, **RIA-Novosti** 167bl, **The Tank Museum** (www.tankmuseum.org.uk) 108t, 109tr, 109b, 114br, 115tl, 133b, 157t and bl, 168l and r, 169l and r, 180l and r, 181b, 188t, 192l, 193t and b, 195b, **Topfoto.co.uk** spine, 83t, 112, 113br, 124t, 155b, 235tr, /Alinari 91t, /Artmedia/HIP 3b, 104t, /Feltz 38t, /Keystone 187r, /Roger Viollet 144r, /Ullstein Bild 37b, 149b, **US Army Signal Corps** (www.history.army.mil) 144l, 203t, **US Naval Historical Center** 4bmr, 9bmr, 33b, 34l and r, 35r, 51t, 52t, 53t, 58t, 59t, 70tl, 76l, 77b, 86t, 86b, 87b, 95r, 125t and b, 126r, 127b, 128b, 129tl and b, 200tr, 204l, 211b, 213b, 214br, 215r, 216t, 221t and m, 233t, 234b, 235tl, 239r.